Economic Optimization of Innovation & Risk

A Theory of Crash Rate for Private or Public Projects

Second Edition

Robert Shuler

Comments on *Economic Optimization of Innovation & Risk* from Gerald J. S. Wilde, father of Risk Homeostasis theory:

"I found your ms very interesting, packed with intriguing and thought-provoking information."

Other books by Robert Shuler:

The Equity Premium Puzzle, Intrinsic
Growth & Monetary Policy`
Special Investor Edition Dec. 2014
2nd Ed. from Tate Publishing coming 2015

Money, Wealth & War
First Edition March 2015
full color paperback – eBook

A Summer Night
The illustrated children's poems
of Pauline S. Shuler
Available everywhere as 8x10 paperback
& in keepsake hardback at Lulu.com

ACKNOWLEDGEMENTS

The author thanks David Pruett for the discussion which began this investigation in 2002. For reading the manuscript and making suggestions also thanks to David, and to Gerald J. S. Wilde, B. G. Smith, David Rutishauser and Scott Askew.

ABOUT THE COVER

The front cover lower portion shows an NTSB photo of the Space Ship Two crash investigation, and the upper portion shows a simulated launch of the Sierra Nevada Dream Chaser. Smaller icons illustrate innovations for which one might want to estimate their crash rate, or how they would affect the risk of other activities, including self-driving cars, ocean drilling for oil, fire safety blankets, tablet cell phones, and cancer screening diagnostic machines.

The back cover accompanies the project example in the last chapter, showing an orbital tourism facility (image based on NASA TransHab) with docked STT (Space Tourism Transport, image based on Dream Chaser).

Cover credits: NTSB, Sierra Nevada,
NASA, Wikimedia commons

Contents

Preface

The seeds of this idea germinated in 2002 during a prolonged and animated discussion between two very experienced project managers at NASA. There was also a contribution from several years of experience in the world of consumer software development and marketing, which had seen two decades of remarkable development and change. The gist of the question was whether tools which made product or system development more productive, and defects easier to find, also made them more reliable.

On the one hand, both agreed that hardware seemed to produce fewer surprises than software. Intuitively this might be due to the fact that hardware was harder to design and change, feature for feature, than software, encouraging more care. On the other hand, an upward spiral of sophisticated tools to aid software development, most of which made it easier and easier, was largely *intended* to make it easier to identify, correct or avoid errors. But would it make software more reliable?

A prime example of this conundrum at the time was the Windows operating system, which had a reputation for crashing and losing user data, but which was tolerated by both consumers and businesses. Windows was developed using a high level language and object oriented methods, and should have been more reliable. One of the participants in the conversation, the author, took the time to write a white paper and derive some equations based on micro-economic considerations.

It was not strictly true that all of the new methods made development easier. Some systems proposed for use by the Department of Defense seemed to have as their object to make software harder to develop and change. By and large these were not widely adopted, however, and one of them, the Ada programming language, was over time a colossal failure due to lack of support and adoption.

From its inception until about the mid-1980s, NASA relied heavily on testing to ensure reliability, at great cost, including manned flybys of the moon which attempted partial descents but not actual landing. Simulators were developed with such high fidelity to reality that the author, when asked to choose between an actual photo of the lunar surface, and the view from a simulator, selected the wrong one. To the extent humanly practical every failure scenario in hardware, software and operations was anticipated and tested, and in the case of manned operations a response was formulated and rehearsed.

While there were spectacular unmanned rocket failures, and the loss of the crew of Apollo 1 on the ground during one of the ambitious and as it turned out ill-advised tests, no crew were lost in space up through the completion of the Apollo program. The robust failure response and simulation infrastructure was able even to cope with catastrophic failure of the Command Module on Apollo 13, and effect a safe return of the crew using an adaptation of the life support system in the Lunar Excursion Module. It was assumed by all, inside and outside NASA, that a significant failure resulting in loss of life would very likely terminate the manned exploration program. While given military importance in procurement matters and in the national budget, at that time, it was never assumed to have the tolerance for loss of life of a military program. It was primarily, if we are to believe historical documents, a program to increase the prestige of the United States with then-called "third world countries" as a strategy in the Cold War, but a civilian strategy held to civilian standards.

After the election of Ronald Reagan, who took office in January of 1983, all government agencies were asked to operate on a basis similar to commercial, for-profit businesses, that is, to be economically motivated. The civilian space program would now be held not only to civilian but also commercial standards. The nearly immediate effect was that in view of pressure to increase flight rate, advice of engineers was overruled and a launch of the Challenger Space Shuttle on a too-cold January day resulted in another spectacular rocket failure – this time with a crew of seven including a civilian teacher, viewed worldwide.

While the manned space program was not immediately canceled, the Air Force withdrew as a Shuttle customer, and the space station suffered a three year delay while a replacement Shuttle was built. The Space Station Freedom program was eventually reorganized by the Clinton administration and major pieces given to our newfound ally, the Russian Federation. It became known as the International Space Station (ISS). A decade followed in which a series of low cost ideas for pushing the manned space program forward (an experiment to make propellant on Mars, a large volume inflatable habitat, a variable thrust plasma engine) were discarded without much thought.

Finally in 2003 another Shuttle catastrophically failed. Once again it failed because of a well-documented and *recurring* problem regarding which many competent engineers had been simply overruled. At that time the author took a second look at the "white paper" and its assumptions, and realized that not only was the economic analysis extensible to very large systems involving both hardware and software, but that it was quite applicable to government systems as well as for-profit systems. They after all do have a cost and are paid for by taxpayers, and eventually are called to account for those costs with results. Government also hires private sector engineers, managers, and management consultants, generally follows the same paradigms as most large corporations, and works with about the same efficiency, though politicians and their industry suitors vying for lucrative contracts like to disparage government performance in their own self-interest.

The extended white paper was circulated within the Johnson Space Center (JSC) with mixed but largely negative reception. Because of the internal politics of the University of Houston system, the local branch offers no engineering degree except System Engineering. Thus many of the upwardly mobile managers and engineers at JSC are influenced by or even wedded to the ideas of System Engineering, which basically are that following a certain methodology – which is vaguely defined in the author's opinion – can reduce costs *and* build systems more easily *and* make systems more reliable. The idea that economic considerations might override that intellectual achievement is anathema when the optimistic group-think of bright people takes hold, people who are told they should do more with less and are intent on not scaling their plans back one bit. However, it is this very blindness which makes them all the more susceptible to the economic realities. They will cut costs until something breaks.

The impact of the Shuttle Columbia's breakup over East Texas in 2003 soon became apparent. The Bush administration set an end date for the Shuttle Program. The fear had come true. An important failure caused the demise of the world's premier manned space program and the only reusable space transportation system. Publically something else was put in its place, the Constellation Program, which proposed to return to the Moon, using it as a base to develop technology that would eventually be used to go to Mars. The manned transportation system would be an old style "safe" capsule, with updated technology to be sure, but from appearances adding little to the nation's prestige as a leading developer and operator of space systems.

Thoughtful members of the space community assumed that the weakly supportable Constellation Program was vulnerable to cancellation, and even during the 2008 presidential primaries candidate Barack Obama was promising to do just that, and use the money for education. It turned out his education initiative waited until early 2015, but the Constellation Program was canceled immediately in 2009.

Seeing the potential applications, and experiencing the consequences of working for an organization that was not able to consciously and rationally manage its catastrophe rate, the author began to seek some opportunity for publication of the crash rate theory. This was a new technical area, and some trouble was encountered selecting the right journal. The general feeling of editors at behavioral economics journals was that there was already a psychological theory of risk compensation or homeostasis, applying to individuals. These were not quantified theories, making it difficult to apply them to a development process, but project development processes were not a concern for journals dealing with the behavior of consumers.

The author then tried a leading system engineering journal and was surprised to find that the editor liked the idea and thought it important. He didn't think the economic content fit within the topic of "system engineering," however, and referred the paper to the editor of a journal dealing with enterprise transformation. The second editor seemed to be interested but had trouble completing the review.

In the meanwhile, evidence mounted for the importance of the theory. The world's most serious truly private spacecraft, Space Ship Two being developed by Richard Branson, Burt Rutan and Paul Allen, endured a crash killing one test pilot, injuring another, and setting the program back years and probably hundreds of millions of dollars. This massive economic jolt will most likely cause fundamental changes in their development and test program. The crash rate equation likely would have suggested those changes were necessary to begin with.

First one Asian airline and then another endured twin crashes within the space of a year, actually within a few months in one case. The reader might wonder what the shoot down of MH 17 over Ukraine has to do with a theory of crash rate. Several things. First, the motivations and actions of the airline in deciding not to alter its flight path (several other airlines had already altered flight paths) was an economic one.   Second, the reaction of the flying public to the twin crashes, the other being the mysterious disappearance of MH 370, was essentially the same as it would be to two clearly "at fault" crashes. The public holds Malaysia Air responsible, and so does the crash rate equation.

So I withdrew the paper from its holding pattern and published in the open access *International Journal of Engineering Innovations and Research*, an Asian publication which encompasses both engineering and innovation.   The title of the paper is "Optimizing Innovation and Calamity," and it is freely accessible online.[1]

In less than a month after publication, the second fatal crash of carrier Trans Asia occurred.   I am convinced that circumstances led to the publication in Asia because that's where the knowledge is most needed.   Generally I am referring to Asia outside of Japan.   The knowledge of quantitative theories of quality management in Japan is extremely mature, more so than in the United States or Europe, and a section of this book will draw detailed comparisons of the crash rate equation to the findings of so-called Six Sigma, which is an American name and not the term the Japanese use for quality management theory and processes.

Several further events prompted the writing of this book.   In the summer of 2013 I developed a presentation for new management and I also filed a New Technology Report, which resulted in the presentation being deposited on the NASA Technology Report Server (NTRS). There it was picked up by an organization called the BilioGov Project, and copies of those charts were made available as a print on demand book for about ten dollars.   I am listed as author.   The title is the same as this book, *Economic Optimization of Innovation and Risk*.   These charts are freely available from the author's website, http://mc1soft.com/papers/ .

However, copies of PowerPoint slides with no explanation are inadequate.   Therefore the present book you are reading, which I have designated a "Second Edition," corrects that problem with 95 pages explaining the material as simply and completely as possible.

Two other events are indirect outcomes of the crash rate theory. Sometime in 2013 I showed the charts to an economist at the University of Mississippi, Robert Van Ness.   He immediately asked if I had tried using the idea to solve the equity premium puzzle.

I had not at that time thought of any connection. But I had previously tried to solve the equity premium puzzle – the puzzle of the un-equalized excess return on equities over bonds discovered by Mehra and Prescott and published in 1985 – as have many other people, and I could see why he thought there was a connection. While I did not use the crash rate equation per se on the puzzle, I did resume work on it and during the three week sequester in the fall of 2013 wrote a book, *The Equity Premium Puzzle, Intrinsic Growth and Monetary Policy*. One of the keys of that book is the role of productivity in triggering unexpected results from monetary policy and in producing the equity premium. We will see later that productivity plays a similar role in triggering unwanted results from the crash rate equation.

The equity premium book led to some discoveries about the role of productivity, and also trade based on labor costs (which in the short term has an effect similar to increasing productivity), in the conflicts of nations. Over the last year I have written a second book, *Money, Wealth & War* detailing those findings. In a way, all of the findings in those two books trace back to that discussion in 2002 and to the crash rate theory. War can be viewed as a "crash" in the ship of state. The paper mentioned earlier attempts to justify such an extension.

If justified, and I think it is, why not extend the theory across the ages to predict the rise and fall of civilizations? Perhaps one of the readers will make this contribution, and produce the real life counterpart of Asimov's famous fictional theory of Psychohistory. I no longer believe that such a theory is so farfetched.

Approaches to Understanding Risk

The Asipu 3200 BCE

The word "history" is itself misleading in this context, as humans almost certainly were analyzing risk in pre-history. The earliest recorded usage of systematic, methodical analysis of risk coincides with the earliest written records in Mesopotamia.

The Asipu were "scholars and practitioners of diagnosis" living in the Tigris-Euphrates valley about 3200 B.C., in the ancient civilization of Sumer. They also functioned as advisers on risky, or uncertain, ventures. They did not claim to be able to see the future or to use divination as we might expect from Greek mythology or the folklore of primitive peoples. Rather, they proceeded by "identifying important dimensions of the problem, considering alternatives and collecting data."[2] The Asipu method was similar to modern risk analysis. They formed a ledger for each decision alternative. They would collect data on the possible outcomes and denote whether each was favorable or unfavorable.

Now here comes the problem. How could a determination be made as to whether a favorable outcome was indeed likely? Modern risk analysts still resort to methods of arcane tabulation which merely obscure the fact that likelihood estimates are guesses made by summing and merging low-level data from many individual analysts or engineers, some of which are reliability data but some of which also are guesses. The Asipu were perhaps more straightforward, simply assigning an estimate of whether "the gods" were favorable to each outcome.[3] Based on post-facto analyses of the "predictions" of modern risk analysis after major failure events, the Asipu may not have fared much worse.

The theory of such a method, modern or ancient, seems to derive from the idea that the average veracity of many small guesses will exceed that of one very large guess. There probably is some merit to such an idea, but the trouble is that expectation biases creep into organizations, affecting the entire spreadsheet with all its little guesses and judgments. This expectation bias then is also measured by the analysis process, not only the risk itself. The expectation bias becomes a new and independent state variable. From time to time it may be inherited from predecessors, influenced by management, moved up or down by various consultants, and corrected by occasional real world events. Unfortunately in no case do we have a real life Mr. Spock in the background, ready with "scientific" probability estimates accurate to several decimal places.

It is the periodic correction by actual events that forms the first basis for crash rate theory. Since the entire edifice of risk analysis is more directly a reflection of this "expectation" state variable than of the often unknowable real risk, it is adjusted like an intelligent control system with two sets of inputs. One is the anticipatory input inherited from various sources, ranging from an analysts' college professors, to management, to the opinions of regulators. The other is the correction signal which occurs when crashes are more frequent and costly than expected.

The second basis of crash rate theory is that the correction signal will be an economic cost, able to be compared and balanced with other economic costs.

Risk Analysis

Risk analysis is based on the hypothesis that statistical and engineering methods can estimate risks and guide hedging. It is considered to be 5000 years old, beginning with the Asipu. It seeks to answer such questions as how many years supply of grain should be stored as a hedge against drought. There are Biblical accounts of how civilizations rose or fell on such questions, though those accounts do not show a systematic approach to determining the answer.

Cistern in India[4]

Cisterns for storing water beg the same questions. How much is enough? Records of rainfall and statistical analysis can answer the question of how much is enough to a certain probability, in a certain time interval. Engineering methods can build the required buffer. As is well known, long time intervals can produce variations that exceed such estimates.

While we think of cisterns as ancient devices, in reality some urban environments such as San Antonio use what is in effect a natural cistern system, in that case the Edwards Aquifer. Many mountain communities use dammed up valleys to store snowmelt to provide year round water access, such as Lake Dillon in Summit County Colorado. One could even say the city of Los Angeles depends on several giant cistern-like structures, including Lake Mead behind Hoover Dam. Sizing such structures, or conversely limiting the size of the dependent communities, is dependent on the methods of risk analysis. The penalty for a serious mistake is a serious water shortage. In the case of isolated ancient cities which were mysteriously abandoned at some point, the inadequacy of cisterns is a leading theory of cause.

Whenever statistical data are available, they tend to be used. Sometimes this falls within the province of engineering data and analysis, as for example the records of strengths of materials under certain stresses. The user of risk analysis must anticipate stresses based on usage profiles, the design of the system, the component and material data available, and the statistical repeatability of its manufacture.

Since these criteria include both human factors and environmental factors, their validity are often no better than the Asipu guesses. Human factors may change. A key failing of most risk analysis is to assume that human factors do not change, and are not influenced by the performance and features of the system itself. A key assumption of crash rate theory will be that human factors in fact are influenced by the features and performance of the system. Financial risk analysis suffers the same two weaknesses, human and environmental (external) factors. Combining a very precise engineering or statistical risk factor with a very vague human or environmental factor produces a very vague factor. Engineers and analysts tend to spend their time on what they can do something about, the well-defined and refinable factors, not the vague ones.

Risk Management

Risk management is based on a management hypothesis, i.e. a call to more complex action than simple drought hedging, and was pioneered by de Meer, Pascal, Fermat, Bernoulli, de Moivre and Bayes from the 17th century. Its tools include insurance, futures and derivatives, for example, the pre-sale of farm crops.

There have been many important modern additions, particularly that of Markowitz[5] regarding the optimal selection of portfolios. In general the concept of portfolio diversification is similar to redundancy. Both can be tools of risk management.

Diversification of suppliers was often used in the early days of high risk aircraft and spacecraft development. It is being used today for the NASA Commercial Crew and Cargo program where multiple contractors are retained well into the development phase. The tools of insurance and futures are widely used by corporations but not generally available to publically funded activities in the U.S., by policy choice.

The importance of risk management to ordinary processes in our economy cannot be overstated, and often the exact terminology is not in evidence. For example, do you call what farmer's do "risk management?" Probably not, yet most American farmers pre-sell their crops on futures exchanges. "Futures" are a tool of risk management, to hedge against unexpected price variations. When visiting Ukraine several times in the 2009-2010 time frame I was surprised by the large amount of unused land, and by how much land use was just individual gardens. I asked my guide and was told it was because Ukrainian farmers did not trust the market to provide a good price at harvest time. Apparently American farmers do not either, and are motivated instead to sell on the futures market, obtaining a known price before they even plant. Southwest Airlines has in some years made record profits because it was more effective than other airlines at hedging fuel costs.

Nearly every American industry which uses large amounts of a market commodity also engages in risk management via various derivative financial instruments based on that commodity. That includes financial hedging on financial markets themselves. Without risk management, much of the productivity of our economy would likely be lost, because facing uncertainty, companies would not invest in supplying the full potential consumption of the markets.

Risk Compensation

Peltzman in 1975 put forward an argument that users of products or systems (such as automobile safety devices) may engage in "risk compensation," maximizing some other utility when risk decreases, so that overall risk does not diminish from the use of safety devices as much as expected, if at all.[6] The result of this is that, for example, the reduction in automobile fatalities from wearing seat belts might be only a fraction of what engineering tests suggest.

Risk Homeostasis

Wilde in 1982 argued for a stronger theory of risk homeostasis (tendency toward a relatively stable equilibrium) in which compensation is more of psychological than economic origin, and individuals adapt their behavior to maintain a target level of risk they are comfortable with, as if they had a risk thermostat.[7]

The literature abounds with studies that both support and refute each of the latter two theories. For a good summary see Hedlund [2000].[8] The general consensus is that they explain at least some of the data, but it is unclear ahead of time when such effects are going to be important. Researchers consider factors such as whether the measures require user action (seat belts) or not (airbags), whether they provide some feel of increased safety or performance (anti-lock brakes), or whether the user may be possibly unaware of the presence of safety measures (side impact resistant reinforcements).

Stetzer and Hofmann [1996] point out that while these two theories (risk compensation and risk homeostasis) posit the behavior of individuals, almost all studies measure aggregate behavior, which they argue can be troublesome to compare.[9] The plan of this book is to posit an aggregate (usually corporate) point of decision. The method of arriving at that decision will be treated separately. It may be deliberate in consideration of economic or moral factors, or it may be evolutionary in that entities which make the most market-effective decisions come to control more resources naturally. Viscusi [2000] has pointed out that consciously making product risk decisions at a corporate level may be detrimental in and of itself, since jurors award punitive damages in injury cases perversely in proportion to the amount of analysis and value of life used in corporate analysis.[10] Nevertheless, general reliability analyses are bound to be made, and corporate culture and policy will result in de facto decisions.

Do the practitioners know the theory?

The details of risk analysis during the *development* phase often fall to *engineering* groups, particularly system engineering, or if it is primarily a software project then software engineering. The author searched and could not find journal articles in the field of system engineering which mention either risk compensation or risk homeostasis. Most of the theoretical work on risk compensation is published in various economics and safety journals, and risk homeostasis was introduced in risk and psychology journals. Each discipline has its own practitioners and preferred viewpoints and approaches.

It is not clear, for example, that system engineers are likely to encounter papers in economics journals, however related to system engineering they may be, nor that they would be impressed by the arguments used, such as the psychology of risk homeostasis. Economists, on the other hand, find such arguments not even particularly novel, since one of the central tenets of economics – modern portfolio theory – takes as its *goal* that investors wish to maximize return at a *given fixed level of risk*. That sounds a lot like what is taken as a postulate to be tested in risk homeostasis theory. And to an engineer it may sound like a superstition to be refuted. In engineering there is a psychological motivation to believe that the tools of engineering work, i.e. that improving the reliability or operability of a system has real benefit and is not optimized away on a whim. Thus the domain of the engineer is *reliability theory*, which examines failures due to the physical characteristics of a system, its environment, and usage.

The data papers examining these theories (risk compensation and risk homeostasis) frequently appear in safety journals. These data are rarely able to distinguish why an effect occurred, but most results confirm some degree of risk compensation or risk homeostasis, for *some* reason. And some of the data are surprising, and strikingly counter-intuitive.

An engineer at the author's institution was involved in a project to improve fire safety blankets in 2002, and when asked about possible risk compensation effects he reacted with understandable disbelief. But the very next year Schindler [2003] reported that effective fire safety blankets, largely used only in the U.S., induce risk taking and a higher firefighter death rate.[11]

It is common for engineers, and even some economists, to calculate costs and benefits of safety measures as if usage and human behavior were an unalterable constant, for example Katarelos [2008] develops a delta-cost/delta-risk model for shipping and also applies it to air transportation.[12] There is no delta-behavior in his model. But in general data suggest that usage patterns change when risk changes. This change in behavior is one source of unexpected outcomes.

The basis for an economic model

The temptation for an engineer to try and develop a quantitative model of such consistent and sometimes dramatic effects is strong. At least some of the data suggests a relation to economic cost models. Parry [2004] holds that tax costs can reduce traffic accidents.[13] Cummins [1999][14] and Cohen [2003][15] report that no fault insurance and mandatory insurance induce unsafe driving. The data are not uniform, and a model might provide *advance* insight into which types of safety or reliability improvements are more effective when deployed. For example, Carlsson [2004] asks if safety is more valuable in the air, since we seem to pay more for it there.[16] Levitt [2001] reports that when sample selection is accounted for, the cost per life saved for airbags is 60 times greater than for seat belts.[17]

While much of the data and theory focuses on the behavior of individual users or groups of individuals, corporations determine much of the risk landscape from which users choose. Corporate behavior, closely coupled to profits and losses, may be more amenable to an economic model than user behavior. The approach we will take is to form an economic model for both corporate and user behavior based on competition, and to analyze how user behavior might deviate when users are not working or competing.

The relation between innovation and operational failure rate has been little studied. Searches for papers relating to innovation and risk reveal much focus on managing the development risk of innovation, but not the operational risk. Searches relating to innovation and safety turn up many studies of innovation for the sake of safety, or innovative safety measures, such as for example the fire safety blankets discussed above, but no study of how innovation generally affects safety or failure rates. Yet that there is such a relationship is well known anecdotally. For example, many computer users habitually avoid "release 1.0" of anything, assuming the bugs have not been worked out. However, in the longer term it is generally assumed that once mature, innovative technologies will provide more reliable operational service. Few would disagree that transportation and communication, for example, are more reliable than 100 or 1000 years ago. Later we will see that a variety of factors contribute to this effect, including the cost of the technology as it becomes a commodity, and changes in the standard of living and length of life which affect the value of life.

However, if users respond to an increase in the safety or reliability of a product or service by adjusting their usage such that the aggregate safety changes, then would we not suppose that they might respond to some other feature in a similar way? In the narrow theory of risk homeostasis, where risk is supposed to be the controlling parameter, maybe they would not. But in an economic model, where all features and also reliability are modeled in economic terms, then once given an economic weight, we would suppose a similar response from users. In this way, innovation in the broad sense becomes entangled with aggregate failure rates.

Why do we specify aggregate failure rates? We do so because this is directly related to aggregate corporate or social costs. Such a model is independent of whether failures are per mile, per hour, per transaction, etc. A general model can be developed, independent of the unique characteristics of a particular technology or application.

Summary

We have seen that risk analysis is at least as old as civilization, and that civilization depends upon it to a very great degree. At its root, risk analysis invariably must combine some precise and some nearly unknowable factors.

The more sophisticated modern practice of risk management involves mathematically complex futures and derivatives, and to a large degree the productivity of civilization depends on this discipline. In some contexts "risk management" is used to mean the management and adjustment of components in a risk analysis process, or even more commonly the management of quality, sometimes known as "quality management." In this book we will prefer the more specific term of quality management where it is appropriate, and use risk management as specified above.

Either risk compensation or risk homeostasis can produce surprising outcomes, the subject of the next chapter.

Unexpected Outcomes

In this chapter we consider characteristics and possible explanations for unexpected outcomes when intervening in risk processes, with a view toward providing evidence for the mostly economic assumptions we'll adopt later. Not all risk processes can be explained or characterized in a particular way, but we'll see that a great many can.

There are two difficulties with discussing "unexpected outcomes," especially where Americans are involved, because of two prejudices:

1. If an analysis is complete, no outcome is truly unexpected.
2. When a sufficiently large force is exerted in a certain direction, progress *will be* in that direction.

The first concern is addressed by being careful with definitions. We mean, obviously, the outcome that would reasonably have been expected based on earlier analysis, or intuition if that is all that was earlier available.

The second concern is more difficult. It is analogous to a belief in Newton's law of motion $F=ma$. If a large enough force is exerted in a certain direction, then inertia, friction and any other factors will be overcome and acceleration will be in the direction of the force. But this may be a very wasteful way of achieving results, and less forceful interventions allow the risk system room to compensate.

Characteristics of feedback systems

Take the case of any system with feedback, which acts like a control system, whether it be manmade or naturally occurring. A good example is the temperature in your home, which I assume to be controlled by a thermostat and an active heating and cooling mechanism. If it is too warm, you may momentarily cool at least some part of the house by opening the refrigerator door. Or you may heat it by using a space heater. Soon enough, however, the thermostat will kick in and other parts of the house, away from where you are applying a forced temperature variation, will compensate with the effect of moving the temperature of most of the house in the *opposite* direction.

It is true that if you apply enough heat, or cold, the home's built-in heating and cooling system will be overwhelmed, and eventually destroyed by trying to oppose a superior force and operating at 100% of its capacity. We assume that no one wants this sort of outcome, unless one is dealing with a criminal subset of society which must be constrained at any cost.

Such systems as the home with a rogue heating or cooling source are not difficult to analyze, and we assume most people would not make this mistake in their homes. Or would they? A good counter-example is the dual-zone temperature controls becoming more common in the U.S. and already standard on most automobiles in Japan. The system on a 2004 model Cadillac was known to enter a mode where one side would go to max heat and the other to max cool, creating an intolerable environment at a time when the vehicle operator should be paying attention to driving. Even cutting the unit off would not immediately solve the problem because the offending temperature has persistence, and it might be necessary to run the unit if it is quite hot or cold outside. In a word, the unit was dangerous, and dealers could not fix the problem. It was intrinsic to the design. While it is possible to design a better dual-zone unit, it is more a matter of luck than theoretical analysis, and I operate mine with little or no temperature differential. But I am an engineer with extensive experience in control systems design and analysis, not a typical consumer.

To a consumer believing the dual controls are meant to be used as such, a temperature war within a vehicle is an unexpected result of attempting to independently control passenger and driver comfort.

Explaining motorway death rates

The last two risk theories we discussed, risk compensation and risk homeostasis, amount to control systems with the potential for unexpected outcomes. Risk compensation is a qualitative theory, and used mostly to "explain" counter-intuitive results after the fact. Risk homeostasis is quantitative in that it holds that individuals manage their total risk to be some target value. This is a bit simplistic, and so it is unsurprising that experimental data hold it to be true only some of the time. We can suggest, however, how it might explain some bizarre results.

"Reasonable & Prudent" Montana Speed Limit

Those of you who may have read my other books will note that I am fond of re-using some of the same examples. I apologize for the repetition, but they are extremely relevant to our current discussion, and in fact we will begin herein to obtain more quantitative insight into how they come about.

Consider the case of the Montana speed limit. When the nationwide speed limit of 55 mph was repealed, Montana for a time reverted to a daytime unlimited speed on certain roads. The accident and fatality rates dropped. When a 75 mph limit was imposed, the rates rose again and eventually doubled. That is statistically significant.

If we try to analyze this only in terms of driving safety, we do not get very far. There is basically no way that we can find that higher speed driving, potentially unlimited speed driving, is "safer" given equal care and attentiveness. We are left with the seemingly untenable proposition that drivers have more fatal accidents when driving slower. Even risk homeostasis does not explain such a situation given these facts alone. We might suppose the slow drivers are bored and inattentive, but their risk of an accident actually rises, and risk homeostasis suggests total risk should remain the same. Risk compensation might posit that the drivers were able to make some other use of the time, perhaps reading or watching TV or talking on the phone while driving slower, and the utility of that activity compensated them for the increased risk. But this, too, is a pretty thin explanation.

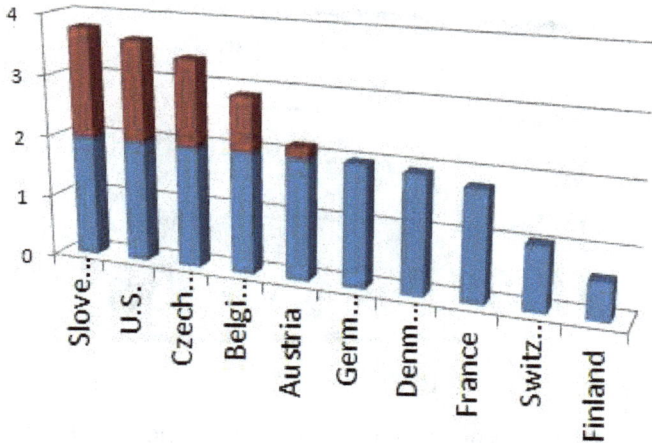

Motorway deaths per billion vehicle kilometers[18]

It might be possible to concoct some explanation peculiar to the culture of Montana, but international data suggest not only that there is nothing peculiar about Montana, but that death rates decline with rising levels of some types of hazards generally. In the chart above we see that the U.S., with usually safe road construction, nearly leads the developed world in death rate, while Germany with unlimited speed Autobahns has half the deaths per mile. Switzerland and Finland with dangerous mountain and winter driving conditions have dramatically lower fatality rates. Any cultural explanation runs afoul of the Montana data.

Here we have a situation in which users are operating technological equipment (automobiles) with some cost and feature set in a complex environment, and the more dangerous we make it, the less often they crash. This is certainly something we want to understand if we are going to make a theory of crash rate.

There is a very simple factor we have overlooked which is perfectly correlated with crash rates in the above examples. That is the cost of the crashes, cost being the extended economic cost, with some relatively high value assigned to loss of life. The value of life in the U.S. ranges from around one million dollars for insurance and liability purposes (20 years at $50,000 per year) to around $9 million for environmental protection analyses.[19] In Russia the range is $90,000 to $4 million. We don't need to assign an exact value for the present discussion, only to admit there is some value, and assume this enters into people's thinking directly or indirectly.

Let's consider what speed is actually safe by "brute force." According to slower-speeds.org.uk that would be 20 mph[20], at which the survival rate is 95% even for pedestrians. Cars are only crash tested in most countries at 40 mph because the integrity of the car's frame doesn't hold above that. Many crash situations that develop at higher speeds are reduced by braking before the actual crash occurs. Nevertheless, brute force safe speeds are not generally tolerated by the citizens of most countries.

A simple calculus shows that the collective additional driving time at minimum wage, due to driving faster, exceeds the economic value of additional lives lost. A typical annual death toll from motor vehicle accidents in the U.S. is 33,000.[21] If we use the insurance value of life of $1 million, that is $33 billion dollars equivalent loss, or about $300 billion using the highest estimate of $9 million per life, the one used by the EPA. Most of the country's 300 million people will spend some time in a vehicle, so let's compute the loss per person. It is $110 using the low value, and $1000 using the highest estimate for each life's value. At a time value of $10 an hour, which is already the minimum wage in some states, eliminating 100% of that loss of life would be worth something like 11 to 100 hours per person per year extra time spent when traveling by automobile. For that cost, for those extra 11 to 100 hours, assuming the average person spends 479 hours a year in an automobile,[22] we would need to reduce speeds by 2.3% to 21%. But for 2.3% reduction in speed it is unlikely we'd save very many of the 33,000 lives. Even at the high end of 21%, that is only a reduction of 75 mph to 60 mph, not all the way back to the once national 55 mph speed limit which was abandoned.

The U.S. annual automobile death toll peaked at 54,000 in 1973,[23] the year the 55 mph speed limit was enacted, and fell to 45,000 the next year, apparently saving 9,000 lives, less than one third of what is required to economically justify the driving delay using the highest estimate of life's value.

I apologize for seeming cold. It is not that I think such high death rates are acceptable, not at all. It is that any theory of crash rate should be realistic regarding what methods are effective in addressing the problem. What we will find is that certain types of approaches are more effective than others, that there are cost-multipliers in certain approaches, but I am getting ahead of myself. The 55 mph law failed to achieve anywhere near economically justifiable savings, the public protested, and the law was repealed. Speed monitoring cameras were shown to reduce fatalities in parts of London [Ibid. 20], a normal not an unexpected outcome, but voters in the U.S. have ousted traffic monitoring cameras in several parts of the U.S., including the Houston area where I live.

Returning to the puzzle of the international data on motorway deaths, consider again the cost of a crash, any crash, in the conditions typical in the various countries. Let whether the crash is likely to be fatal be part of the cost. A crash at high speed is much more likely to be fatal. All sources agree on that point. Now we must deal with a further complexity. Whose speed?

That's right, it is not just your speed, but the general speed of all traffic that is considered. The Montana country highway in the late 1990s and the German autobahn have a high cost per crash because other cars are going very fast, and any collision is likely to be fatal. We can calculate the cost of crashes to an individual as the expected cost, which is the probability of a fatality times the cost of a fatality. Reducing one's own speed below that of other traffic while attempting to occupy the fast lane of the autobahn is, I would think, inviting a higher probability of a crash. One must drive with other traffic.

Even if there is little other traffic, one must not present an unexpected hazard. And regardless of traffic flow considerations, one finds that one's risk has increased because the crash is more likely to be fatal. The rational reaction is to pay closer attention to driving and maintenance and vehicle selection so as to have fewer crashes.

Here is where the problem becomes quasi-paradoxical. We are assuming in the case of Montana or Germany a higher average speed, which is a given, and is not changed by your personal speed selection. You may indeed choose to cope by driving slower if you can find a lane in which you can reasonably do so. However you do it, the increased cost of a single crash will, we suppose, induce you to reduce the number of crashes.

Since the presumed average speed is unchanged by this "personal" calculus, then the percentage of crashes resulting in fatalities is not changed by it either, so *the total number of crashes and therefore the total number of fatal crashes will also be reduced.*

If you can understand that calculus, you can understand crash rate theory.

Finnish highway – safest in the world?

Consider the logic of extending this to driving conditions in Finland. It is much easier to grasp because unlike the situation of speed, the danger factor is in large measure in the environment, and we would not expect it to change in response to human decisions. It is just "there," take it or leave it. Another factor is the commonness of the danger. It is not something novel. Finnish drivers face the hazard a large part of each year. Without doubt you or I would immediately be involved in a wreck under such conditions, as I have been more than once when visiting Colorado which has similar conditions. But the Finnish have adapted and reached a static equilibrium. The concept of static equilibrium will be important later in our derivations.

The Finnish drivers face a high probability of death from crashes, even low speed crashes. If they become stuck on a lonely road in bad weather, they will probably freeze to death. So they will be very careful not to have any kind of a crash. If they do have a minor crash and freeze to death, the cause of death is likely to be attributed to weather rather than the crash, and so the statistics for Finland may even appear artificially low. In any case, without the circular loop of speed choice in the equation, and without the learning curve of unfamiliar conditions, the logic of Finnish motorway safety rates is clear. Obviously these roads would not be safe if used in the same manner as the German Autobahn, and just as obviously the Finnish do not use them that way.

Evaluating the motorway explanation

What factors have we identified that contribute to the cost side of the analysis so far?

- Value of a life, which varies by country & culture
- Probability of death in case of accident, which increases with speed and inclement environmental conditions, and decreases with improvements in road infrastructure and vehicle choice or design
- Probability of an accident, which decreases with improved roads (we presume) and has an unknown relation to speed (reduced fatigue and increased attentiveness vs. greater difficulty of vehicle control)

The factors so far are uniform within a country except for individual choice of driving behavior. But individual choice is an open door to a broad set of conditions, such as the rate of drinking and driving, or even talking on the cell phone or eating while driving. Car & Driver's German correspondent Jens Meiners noted that *"nobody is ballsy enough to try to text and eat a hamburger at 120 mph."*[24]

Have you personally ever driven 120 mph? I have, but only once. It was on a wide, straight and empty stretch of interstate highway, just a couple of years after I acquired a Jaguar XJR. It is very comfortable to drive at 90 mph or even 100 mph. I found the experience of driving 120 mph required too much concentration. The slightest deflection of the steering would jerk the car in one direction or another, and I agree with Meiners. Those not attentive at this speed are immediately removed from the gene pool. Of about half a dozen accidents in my life, three happened when I was driving moderately but distracted, all when I was young. The remaining three happened at speeds of 30 mph (two of them) and zero. I have found that taking an occasional defensive driving class does in fact lower my "near miss" rate. So my anecdotal conclusion is that "attentiveness" and related driver choices can make more difference than speed, and could explain at least some of the data.

If when the probability of a crash turning into a fatality doubles, the number of total accidents is halved, then the data would exactly fit a risk homeostasis hypothesis, in which a "risk thermostat" keeps total risk exactly constant. Direct data are very hard to come by. I spent considerable time searching for non-fatal accident data in Germany. Finally I found the results of a telephone survey which did not break out automobile accidents, but noted: *"a nationwide comprehensive recording of injuries caused by accidents does not exist."*[25]

We can indirectly get at it another way. We know approximately the relation between speed and the chances of survival from statistics like those cited earlier. [Ibid. 20] We can then infer the non-fatal accident rate from the average speeds driven in a country. Again we are stumped by lack of hard data. At least one driver who has tried the Autobahn reports that *"traffic can be heavy enough to restrict speeds to little above the typical motorway speeds found elsewhere."*[26] Also speed limits apply at intersections and construction sites, which are generally regarded as the locations of many accidents. Several sources mention reduced fatigue as a contributing factor, but that would be true only over a fixed travel distance. If motorists simply drive the maximum distance practical in a day, then fatigue should not be a factor, though attentiveness vs. boredom remains a possibility.

The Economist attempts to explain the deliberate lowering of death rates in Sweden,[27] which presumably has driving conditions, values and other factors similar to Finland. The use of "economics" in the journal title would suggest that their analysis might be economic in nature, but they do not describe a direct calculus of the value of lives vs. commute time. The article does emphasize a deliberately high valuation of life by Sweden. They consider no fatalities acceptable. But they also claim to value mobility, so while they use low speed limits they are selective about it, and they use many other techniques, and appear to be succeeding. It would seem that we might learn from them something about the hidden risk control system (or thermostat, as Wilde calls it) which would be valuable in a crash rate theory.

First the article notes that death rates are lower in rich countries. There might be a tendency to suppose that this is because money makes roads better, except for the high death rates in the United States. What it definitely does is make lives more valuable. The cost of a life in a wrongful death lawsuit is sometimes computed directly from the unrealized earning power of the deceased individual. *The Economist* notes that the Dominican Republic has one of the highest motorway death rates in the world. An article from just six months ago in *Dominican Today* documents that the average wage there is not even enough to eat.[28] In fact it covers only 53% of the family shopping basket which 90% of the households consume.

I do not understand how people live in the Dominican Republic. Maybe they have gardens like Ukrainians and some Russians. But it clearly correlates with our developing theory of motorway deaths, and the Dominican data were discovered after formulation, while looking for confirmation. The average wage of 14,279 pesos works out to about $320 U.S. We might expect death rates there to be more than 100 times those in the U.S. from that data alone. In a recent WHO report[29] the Dominican Republic lists 41.7 deaths per 100,000, less than four times the U.S. rate of 11.6. However, not so many citizens have money left over to buy a car after buying food, so the deaths per vehicle are 151.5 or 11 times the per vehicle U.S. rate.

Now if we only had the average wage of the drivers of vehicles we could perform a full test of the theory. We do not, but the number is in the right ballpark. If the average vehicle owner only makes five times the amount necessary for eating, the numbers are *exactly* right. The Dominican Republic is nowhere near the worst, by the way. Several countries have death rates per vehicle 100 to 1000 times the U.S. rate. It would seem the only way to achieve such rates is if drivers routinely run down pedestrians for sport or for quasi-military reasons, which is another matter, beyond the scope of this book.

Returning to Sweden for our wrap-up, there is one more very important insight to be gained. A key to their efforts comes under the heading of policing, or perhaps we should say "testing." In another industry this might be called "inspection." Testing for alcohol, for example, is frequent, and now less than a quarter of a percent of drivers tested are over the legal alcohol limit. The article also mentions vehicles with built-in breathalyzers, a high tech and low cost way of identifying a small percentage of drivers who are by their preferences causing a large percent of problems, and a way of making the cost of driving intoxicated for them very high (i.e. denial of driving privileges).

The author was recently in a store called Car Toys for the first time, awaiting the installation of a new radio in the now very old Jag. Several customers came in to have their court-ordered breathalyzers recalibrated. This "testing" process fits right in with our theory developed back in 2002, and with the older Japanese idea of Six Sigma.

In summary, the *evaluation* of a quantitative theory of crash rate – and here we took risk homeostasis as an example – proves to be very difficult. There are too many variables and the data are too hard or too expensive to get. One way of dealing with this is to "calibrate" the model based on an operating point, and to treat it as an incremental deviation model.

Driving in Ukraine often involved passing unusual traffic, or driving three abreast on a two lane road. Drivers responded to complaints saying "I want to live too." Smashed cars on pedestals were posted as warnings, and pedestrians were passed with clearances less than an inch. The author did not drive there, but did in Russia where conditions are more like those in the U.S. Although Ukraine is more scary to an American, somewhat like videos you see of driving in Italy, the accident rate is only marginally higher than in the U.S., whereas in Russia it is twice as high. Again vehicle ownership distorts the picture. A good many Russians own cars. The per vehicle accident rates are about 2.5 times the U.S. rate for Ukraine, and 4 times the U.S. rate for Russia, so perhaps the author's choice of where to drive was based more on familiarity of road conditions and driving styles than actual safety. A one-time decision must necessarily be viewed as not being in static economic equilibrium. (author's photo)

Wilde's defense of risk homeostasis

Wilde aggressively defends his theory of risk homeostasis, and provides many data examples in a recent book chapter.[30] A figure from this chapter portrays driver adaption of risk behavior as a complex feedback control system, shown below.

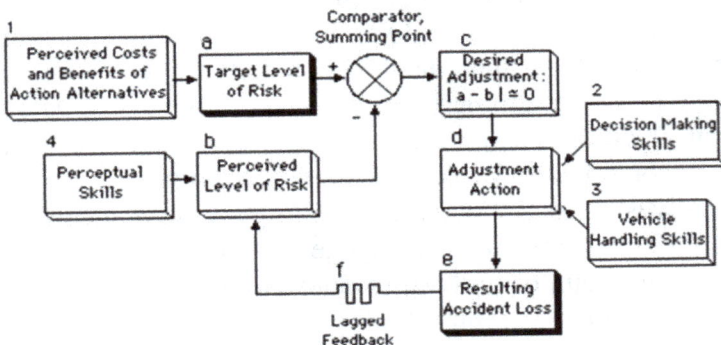

Homeostatic model of accident rate vs road-user behavior with average target level of risk as the controlling variable

Wilde presents data that refute the effectiveness of Anti-lock Braking Systems (ABS), motorcycle helmets, advanced driver training (to increase skills, not just increase perceptions of risk), pedestrian crosswalks, even conspicuous enforcement of blood alcohol levels. The idea so frustrates researchers and safety advocates as to generate comments like: *"Wilde's law of the conservation of misery."*

Crash rate theory will also involve a feedback loop which is capable of producing unexpected effects. For now, notice that risk homeostasis, while discussing economic concepts such as utility, is applicable without an economic framework and primarily applies to individuals. The risk thermostat is their internal feeling, and applicable to primitive humans. Crash rate theory is based on rational economic premises and is applicable to corporations run by rational procedures.

Wilde emphasizes that *"...the accident rate can only be influenced by factors that affect the target risk level."* There are two problems with this. It creates resistance among safety experts and engineers who have to believe they can affect outcomes, and who feel powerless to "persuade" individuals to change their target risk level, if such a thing exists. It seems to fail the intuitive test that transportation, communication and many other activities become safer over long periods.

Risk homeostasis specifies only how the target risk level is utilized, not how it is set. The author feels, on that basis, that risk homeostasis is probably a correct and very strong theory, but complete only in regard to how the target risk interacts with the homeostatic (thermostat-like) mechanism to adapt risk taking in conjunction with environmental risk. It is mostly silent upon how the target risk might be updated. Using an economic hypothesis, if the length and quality of life increases over time, then the cost of a certain risk level in terms of lost lifetime rises, explaining long term trends, as well as data like the Dominican Republic's high road fatality rate. It might be that for individuals, some factors affect target risk very quickly. For example, if one perceives a greatly increased risk in the environment (danger or hazards), one may immediately engage in risky actions (swerving in traffic, drawing a gun or knife in response to a threat, joining a militia to defend one's home or country).

Other examples of unwanted outcomes

If initially some outcomes are unexpected and we become more sophisticated in predicting them, then they become merely unwanted. These remain a problem because they are not a straightforward result of applying force, or "a bigger hammer." Or if they are, the result may be very expensive. Without going into the detailed analysis that we did for motorway deaths, let's look at some additional examples to show how common they are, and how people and policymakers persist in programs that are merely too expensive, but occasionally repeal programs that do not have the desired effects.

Levitt & Porter, Rev. of Econ. & Stat. 2001	**Cost per life saved:**
Seat Belts	$30k
Air Bags	$1.8M

Seat belts save lives. We've all heard that and it is true. Perhaps they do not save as many as engineering calculations projected because of some compensation that drivers engage in, but it appears only to be a partial compensation, not a complete homeostasis. The cost per life saved for seatbelts shown in the above table is a fraction of the lifetime earning power of most people in the U.S.

Airbags are a different story. The cost of $1.8 million exceeds by nearly double some of the lower American estimates of the value of life. Airbags are dangerous to short people, and failures of the technology such as shrapnel emanating from deteriorating airbags have proved fatal in cases where the crashes were not serious.

Airbags continue to be required in cars. The cost per life is high. The cost per car is about $500.[31] Though more than the $200 on which the program was originally sold to Congress,[32] it is still less than many frivolous options people add to cars, and keeps no one from buying a new car.

The same cannot be said for repairs, and even less so for older cars. Expect to pay $2000 to $3000 to replace both airbags after a collision, even a small fender bender if it causes them to deploy.[33] In an older car the airbags just fail from age and if you want them fixed it is the same $2000 to $3000. In a 20 year old classic car like my Jag, new bags from the manufacturer are not available and bags from salvage yards probably do not have enough life left in them to warrant their use, so the cost of custom airbags easily exceeds the resale value of the car. In general, one mechanic told me, cars made after the mid-1980s are not able to stay on the road like cars from the 1960s, 50s or even the 20s, because the computers in them cannot be repaired or replaced. This destroys a source of inexpensive vehicles which exaggerates inequality in the standard of living.

On the one hand, this directly increases the cost per crash for even minor crashes, perhaps even enough to justify a reduced accident rate from repair costs alone.

On the other hand, these costs also induce people to take more risks. Someone in the market for an old car likely will drive it without an airbag, which is permitted as far as I can tell from reading my state's regulations. We cannot exactly argue that old cars are less safe than they would be without airbags (assuming it does not strike them with shrapnel), but we can say that the person with the highest accident potential because of their low earning power (implicitly lowering the cost of a fatality) is the very one that the airbag policy leaves unprotected. The driver of a luxury car needs them less, since high fatality cost is a strong deterrent.

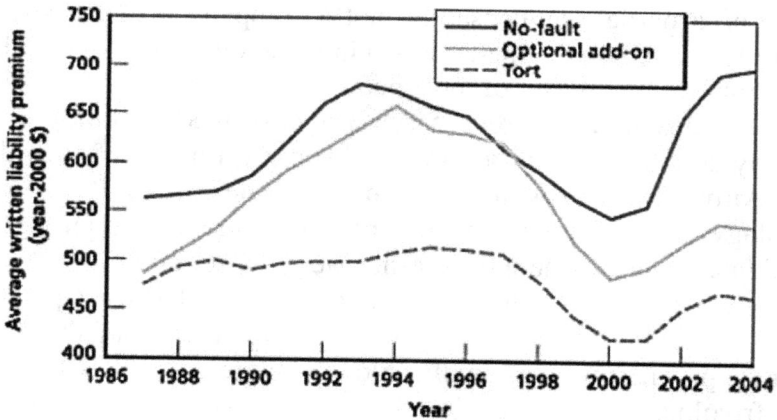

No-fault effect on insurance premiums[34]

Back in the 1980s some accountants at insurance firms figured out that much of the cost of automobile accidents was in litigation, which did not directly benefit accident victims or contribute to repairing or replacing vehicles. The idea of "no-fault" insurance was adopted in some states. Everyone was required to have insurance (already the case) and each person's insurance would pay for their own vehicle repair. As you can see from the preceding chart, it did not work. No-fault premiums grew eventually to nearly double the comparable conventional premiums, and many of the programs have been canceled.

It seems that if accidents have less cost people will have more of them and vice versa. This is exactly the principle we used to explain the motorway statistics, except that the motorway problem was compounded by probabilities of accidents, probabilities of an accident being fatal, earning power of people and value of life. The insurance problem is one of a simple difference in direct costs.

Again we cannot help but note that as a deterrent, such costs are blunt and inefficient. They are not as bad as commute delay costs which apply whether one has an accident or not. But they are not nearly as selective as built-in breathalyzers. Sometimes people have accidents that are just accidents, not because they are bad drivers and likely to have another. The imposition of costs such as escalating premiums has deterred many people from even using their insurance. They will pay for repairs directly if possible, even if greater than their deductible, to avoid years of rate hikes. On an older car, such as the Jag, it makes no sense to carry anything except state minimum liability, because the car is hardly worth more than the deductible. As a result some insurance companies are now advertising first accident forgiveness. Time will tell if that is effective. One thing I can see is that requiring everyone to buy insurance is bound to drive up the cost by creating a compulsory market. But at least we have choices about how much to carry. Think about driving with only the state minimum liability and realize how careful you would be. This could actually lower accident rates.

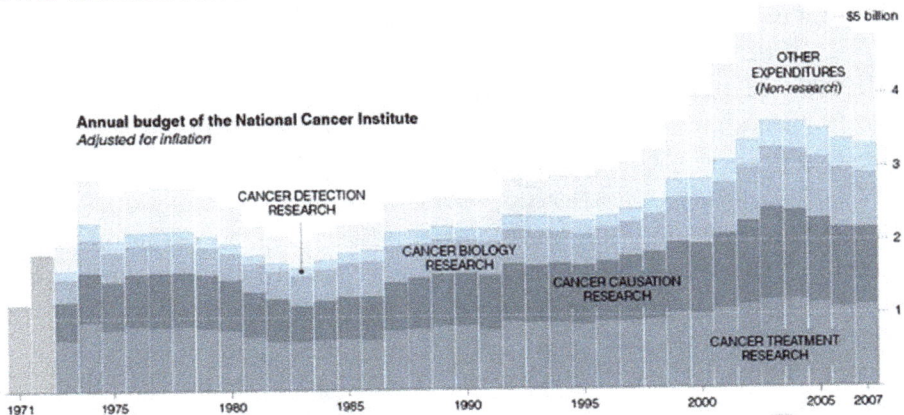

Annual budget of the National Cancer Institute

The budget for the "war on cancer" is shown in the figure above, ranging from 2.5 to 6 billion dollars annually since started by Richard Nixon in 1971, with a cure promised by 1976. Has it worked?

NYT "Advances..." 4/23/2009	Change in death rate or incidence:
Cancer	-5% death rate (since 1970s)
Heart disease	-64% death rate (since 1970s)
Flu & Pneumonia	-58% death rate (since 1970s)
Smoking http://www.infoplease.com/ipa/A0762370.html	-54% incidence (since 1960s)
Illegal drug use http://www.umsl.edu/~keelr/180/trends.html	+6% incidence (since 2002)

Change in death rates or incidence for major health issues

We see from the table above that the decline in cancer death rates of -5% is largely in the statistical noise. It has had only slightly better results than the "war on drugs" which completely failed.

Dying of cancer may not be the kind of "crash" that results from causal factors that we can measure. Certainly some risk factors are known, but one of the best known is smoking and we see from the chart that smoking has decreased by over half.

It is not true that people cannot modify their diet and lifestyle. Not all of the 64% heart disease reduction is from people taking statins to lower cholesterol. New studies suggest that statins may not even help people who don't already have heart disease.[35] The science behind the medicine of heart disease may be little better than that for cancer, but somehow heart disease has been reduced.

In the case of smoking, heavy taxes were placed on cigarettes. Estimates of reduction vary from 4% to 11% for each 10% of tax.[36] Larger Canadian warning labels led to 3% to 5% fewer smokers.[37] These were a low cost measure. Cessation of advertising was also a low cost measure, even saving money for tobacco companies. City ordinances regarding smoke-free spaces also had no economic costs, despite rumors that bars and restaurants would be deserted. You simply cannot keep people from eating and drinking. Ask the highway safety people about that.

The point of these last few examples is mostly that our civilization has a very bad track record at solving problems by the brute force or hammer method, and we should expect crash rate theory to be subtle. A theory of crash rate cannot tell us exactly what to do to fix a given crash rate if knowledge issues are involved, but it might guide us as to the areas in which our methods and knowledge are insufficient.

Development of the Theory

General comments about methodology

The primary purpose of the previous chapter was to acquaint the reader with data which suggest the truth of the assumptions we will make, especially the economic value of crashes. Ultimately the assumptions are just that, assumptions. They are taken as given, and the theory emerges by fairly straightforward mathematical reasoning from them. If the assumptions are not accepted by the reader, no amount of building on them will produce anything the reader cares about.

Since this topic is of interest to multiple disciplines, readers will not have unified expectations regarding methodology. Some will be expecting a statistical approach. This is ruled out by the nature of the problem, which is a small number of large projects that cannot be repeated in a statistical trial (no one would pay for it), and would affect the entire economy if such a trial were undertaken. Without a controlled trial, comparisons between projects of one company vs. another would show nothing regarding a static equilibrium theory, because at any one time those companies would not be in equilibrium. This constraint is common in economics where it is not possible to isolate large scale controlled trials from the general market.

Instead the fundamental and mostly economic assumptions are supported by the extensive historical data already presented, the development of the theory follows careful logic rather than statistical correlation, and future verification will be a long term confidence building process.

Use of linear approximations

The relations we seek are likely to be non-linear and perhaps highly complex. However, at each operating point, for small changes in product features, production process, and usage, it should be valid to approximate the relationships as incrementally linear. This is true provided the relationships do not have extreme sensitivities to small changes, i.e. discontinuities. This will generally involve assuming an incrementally linear relationship, supposing a constant of proportionality, and giving that constant a descriptive name. However, when the operating point changes significantly, the "constants" are no longer constant. Most data-oriented studies suppose only one such relationship, and attempt to discover if it is true. Our methodology will allow a more complex relationship to be developed, which has greater possible explanatory power, but which is more difficult to verify with data.

The straight line in this figure is a linear approximation to the curve in the neighborhood of the point of tangency

Cost of change axiom

We assume that corporations will add features, make other changes, or introduce product lines, which we will call "features," until there is no *incremental* profit from doing so. We assume that if consumers are willing to pay more than the cost of design, test, production, marketing and distribution, then the products (features) will be produced. "Features" also encompass performance.

Simple examples would be the range, speed, passenger capacity and automatic operation of aircraft, or the ability to process merchant transactions easily through the internet. Supersonic passenger planes have gone out of service because the incremental revenue did not cover operation nor did it cover future development. But conventional aircraft range, capacity and automation have been greatly expanded. Range can lead to more and longer flights out of radar coverage over open ocean or hostile territory.

Automation can lead to pilot lack of proficiency and inability to take over in an emergency, even to correct simple landing misalignment. Greater passenger capacity increases the loss of life in event of a crash. Lower costs lead to greater utilization and even to utilization in more marginal conditions.

It is of no concern whether this happens quickly via the shrewd choices of managers and market analysts, or slowly via random experimentation with products. The eventual equilibrium is assumed. All factors of consumer utility (what consumers are willing to pay), and all direct costs are subsumed in the term P_f which is "profit from the features." P_f provides the incremental incentive to add features or otherwise make changes that benefit the corporation.

One term is taken separately, which is the aggregate cost of the operational crash rate of the product, C_R ("cost of the rate"). This is the revenue lost due to a specific crash *rate*. It may be lost due to customers choosing other products, getting bogged down and not being able to buy more product, or imposed costs such as fines, penalties and lawsuit losses. It includes only the costs that the corporation eventually bears in some form, not costs that society bears.

Profit and crash rate cost are treated as aggregates over any convenient time interval, such as a product life cycle, or annually. The cost of change axiom in these terms is stated as

$$P_f - C_R = 0 \tag{1}$$

Keep in mind that both terms are aggregated costs, total profit vs. total cost of the crash rate. The cost of change axiom is inspired by, but not necessarily identical to, the concept of a balance between marginal utility and marginal cost, which is well known in the literature going back to Adam Smith or beyond, and usually stated from the consumer point of view. The marginal utility of safety was noted by Spence [1975[38], via Savage 1999[39]]. Consumer willingness to pay for a marginal increase in safety comes to be equal to the marginal cost of supplying that safety. Safety may be thought of as the inverse of the crash rate. The cost of change axiom posits this from the point of view of costs to the producer of the safety, costs being the negative of utility, and isolates that cost from all the other costs to the producer. But it is really a consequence of the principle of marginal utility, and is adopted as axiomatic.

Development Crash Rate Approximation

We have not yet defined crash rate, specifically "crash." Any consistent definition can be used, as long as it is not so qualitative that it cannot be measured. For example, a computer crash is easy to define. The computer stops working and has to be re-booted. The same definition can be applied to a specific software program. Or one can set a threshold of interest. For automobiles, a dollar value of damage can be set as a threshold, or loss of life, etc. For aircraft, unplanned or uncontrolled contact with the ground will usually suffice, literally a "crash," but even there a damage threshold might have to be set. So for our purposes, a crash is some kind of *failure* of interest, and crash rate may be taken to be equivalent to the usual metric of *failures in time (FIT)* used in reliability theory. Failures in time can be taken reciprocally as *mean time between failure (MTBF)*, useful when estimating the number of hours equipment can be reliably counted upon to operate. For cost and impact purposes, the FIT concept, which we call crash rate, is convenient.

It is traditional (and often required by regulation, not to mention sound engineering practice) to collect information during the development process to estimate operational failure rates. Products, systems or equipment are evaluated during the design phase by analysis. Problems are found and corrected. Prototypes are built and tested, and fixes are developed for unacceptable failures, whether design flaws or issues of reliability and fatigue. Some types of products, such as software, involve failures almost entirely of the design flaw type, but are so complex that not all design flaws can be found and removed from finished products. In either the case of latent design flaws, or materials reliability, the development, testing and certification processes provide indications of the eventual product reliability. For this reason, we wish to have an explicit term for development failures (development crashes) in our model. Since we have an economic model, to predict crash rates, this term will need to consider the cost of the development crash rate.

In the original theory and published paper, it was assumed the costs of development crashes would be the costs of finding and fixing failures during development, testing and certification. These are significant costs. For example, the costs of certifying automobiles for the U.S. market keep some overseas manufacturers from selling cars in the U.S. The cost of certifying airplanes is also very high. For products that consist mostly of software, or firmware, which account for a large share of all technology products, it can be argued that initial design is very easy and most of the costs are for debugging. Certainly testing is a large fraction of the cost of spaceflight equipment and vehicles. It can cost hundreds of thousands of dollars, for example, to certify an inexpensive battery for use in space.

However, much effort in all technology fields goes into tools, processes and procedures designed to make things easier to design, which usually makes it easier to find and fix design flaws. The fields of system engineering and project management make a specialty of identifying problems early to reduce the cost of finding and fixing problems. Likewise reliability engineering attempts to reduce such costs during development, and prevent problems from occurring in the first place. It is a fair statement, then, that most techniques which reduce the cost of developing features also aim to improve the development process and will reduce the development crash rate per feature. In order to reflect this reality, the total development costs of all kinds are used in the present updated crash rate model.

The actual relationship between the development cost of features, and the development crash rate, may be complex and non-linear. But given an operating point, a certain level of features and tools, and assuming the relationship is approximately continuous, we can linearize the relationship as discussed earlier. For this purpose we invent an arbitrary constant of proportionality, K_1, which relates the incremental cost of the features to the development crash rate:

$$\Delta cost(features) = K_1 * crashrate(development)$$
$$\Leftrightarrow K_1 = \Delta cost(features) / crashrate(development)$$

It is apparent that K_1 is the "cost of development crashes," for which we will adopt the symbol C_d. This includes all engineering costs, recurring or non-recurring. To arrive at the total cost of the features, we must add manufacturing, marketing, distribution and other costs which we will designate as M, and we assume they accrue over the product life cycle. These are primarily recurring costs, but not engineering. Using C_f as the marginal "cost of the features," and R_d as the marginal "development crash rate," we can state:

$$C_f \approx C_d R_d + M$$

It will be useful to rewrite this relation as the *development crash rate approximation* giving:

$$R_d \approx (C_f - M) / C_d \qquad (2)$$

You can think of R_d as the essential quality of all technology, materials and methodology used, including performance margins, reliability of materials, etc.

Operational crash rate approximation

The post-deployment failure rate is often found to be related to the developmental failure rate by what is called the defect ratio. In the case of conventional reliability measures, this can be the ratio of defects in the field to defects in a pre-deployment screening process. In the case of complex technology or software products, an analogous quantity is sometimes called defect leakage, i.e. the ratio of defects which "leak" through the testing and certification process. Because of complexity and the randomness of use, these show up over time as the product encounters ever varied conditions of use and/or the environment and technology around it changes. We adopt R_o as the "operational crash rate," and use D for the "defect (or defect leakage) ratio," and assume that a marginal (incremental) relation between development and operational crash rate can be approximated as follows:

$$R_o \approx DR_d \tag{3}$$

If R_d is the essential quality of all technology, materials and processes used during development of a product, then D is the essential quality of integration, test, inspection, maintenance and operations applied after development. Careless operation or lack of inspection and maintenance can undermine any system.

Substituting the development crash rate approximation (2) into the operational crash rate approximation (3) we have:

$$R_o \approx (C_f - M)D / C_d$$

$$\Leftrightarrow C_f \approx R_o \frac{C_d}{D} + M \tag{4}$$

With (4) we now have elucidated an indirect relation between the cost of the features, and the eventual operational crash rate. This relation is enforced by our assumption of eventual equilibrium, and of course will not hold in the transient case as companies make unprofitable "mistakes" in product development. But eventually it will hold, either because corporate managers and engineers are smart enough or lucky enough to find it, or because their competition finds it and supersedes them.

Profit axiom

We now introduce the economic utility to customers of the product with the specified additional or improved features as a fundamental measure, which we equate to what they will pay, and therefore a measure of the revenue the producer realizes from the sale of the product. We can measure this revenue, denoted V_f for "value of the features," per life cycle, per annum, or in whatever way is convenient, as long as it is consistent with the other terms in our model. Using our model of costs (4) we can express the profit derived from this revenue stream as:

$$P_f = V_f - C_f \approx V_f - R_o \frac{C_d}{D} - M \tag{5}$$

The crash rate model

Using (1) to substitute the equilibrium incremental crash rate costs for profits in (5), and solving for crash rate, we have:

$$C_R \approx V_f - R_o \frac{C_d}{D} - M$$

$$\Leftrightarrow R_o \approx (V_f - M - C_R)\frac{D}{C_d} \tag{6}$$

We can further express crash rate costs in terms of crash rate by defining a new term, "average cost per crash" C_c, such that $C_R = C_c R_o$. Using this in (6) and again solving for crash rate we have the crash rate equation:

$$R_o \approx (V_f - M - C_c R_o)\frac{D}{C_d}$$

$$\Leftrightarrow R_o(1 + \frac{C_c D}{C_d}) \approx (V_f - M)\frac{D}{C_d} \tag{7}$$

$$\Leftrightarrow R_o \approx \frac{V_f - M}{C_c + C_d / D}$$

A consolidated reminder of the definitions of terms is provided in the table below (next page). Dimensional (units) analysis verifies consistency. In the numerator we have costs per aggregate unit (per product life cycle, per flight, per year, etc.). In the denominator, both terms resolve to costs per operational crash (direct vs. development). So cost units cancel and the denominator of denominator term "crashes" comes to the numerator, giving "crashes per aggregate unit."

R_o operational crash rate

V_f value of the features (or function) per unit time

M cost of manufacturing, marketing, distribution per unit time

C_d (engineering) costs per development problem

D defect ratio

C_c cost per crash

Terms used in the crash rate model

The figure below provides an interpretation of key terms and how they affect the crash rate, emphasizing those that may have effects slightly different than intuitively expected. I suggest studying the figure for a few minutes before reading the examples. If questions remain after reading the examples, then try reading this chapter again. Following the derivation, and gaining confidence in the formula through its application to examples, are independent ways of approaching the model.

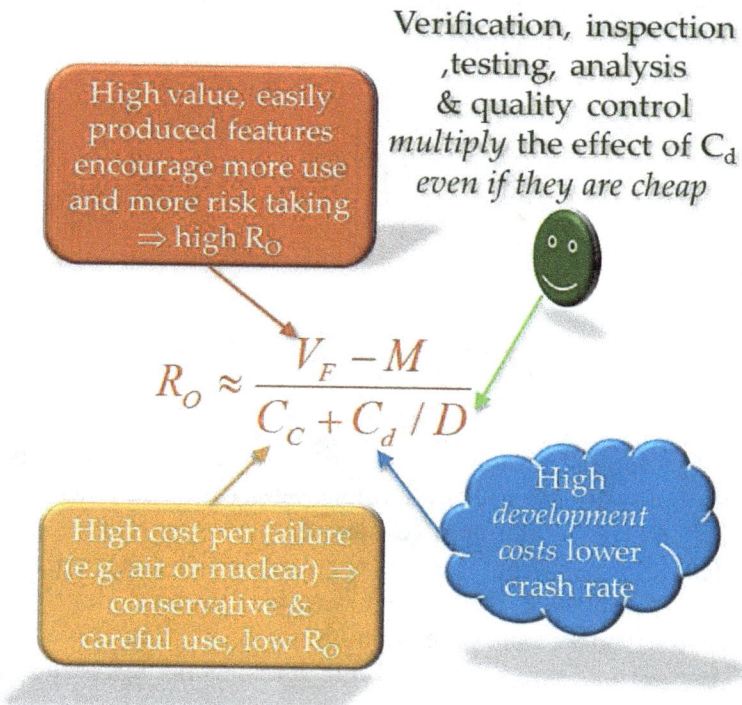

High value, easily produced features encourage more use and more risk taking $\Rightarrow$ high R_O

Verification, inspection ,testing, analysis & quality control *multiply* the effect of C_d *even if they are cheap*

$$R_O \approx \frac{V_F - M}{C_c + C_d / D}$$

High cost per failure (e.g. air or nuclear) $\Rightarrow$ conservative & careful use, low R_O

High *development costs* lower crash rate

Interpretation of the Crash Rate Model Terms

Analysis & Interpretation

While keeping in mind that equation (7) is an incremental relation between marginal costs and rates, not absolute costs and rates, we can nonetheless draw many interesting conclusions. This is especially true for cases where one term or another clearly dominates. We will look at several categories of products to see if experiences are consistent with the model. But first, some additional interpretation is in order regarding the less intuitive model terms.

Development efficiency vs. C_d and D

The original 2002 discussions leading to the development of the crash rate equation, or model, concerned the efficiency of various development processes. We now examine how efficiency is reflected in the model.

Generally testing and the related concept of quality are separate disciplines, with specialized tools and dedicated professionals. For purposes of discussion, we continue to consider them separately, but realistically they are overlapping. Developers attempt to improve their methods to generate more features (designs, lines of code, consumer products, bestselling books or movies, spacecraft, whatever it is that they do). They also attempt to generate higher quality items, with fewer bugs or glitches, either by design or by improved screening before they hand off to validation, testing, quality, marketing, or whatever the particular industry calls the step following design in which the product or service is made actually ready for consumption.

The testers, under which name we subsume all those things we just mentioned from validation through marketing, might occasionally suggest a design change but are by and large focused on identifying problems.

If designers use a new tool, process or method that increases their productivity, say for example enabling 4 times as many lines of code to be produced, generally there is an implicit assumption about the error rate of that code. First of all there is an error rate, always. Second, if there were no constraint on the error rate, the range of productivity already is so variable – from one line of code per day for flight software to 1000 per day for non-critical administrative, training or data retrieval and display software – that a small improvement like "4" is lost in the variability.

For that reason, it is irrelevant in a crash rate model to have separate terms for productivity and for the error rate of the production. The term C_d combines the two. Since it is a "cost" term it is effectively the inverse of productivity. If for example our new productivity tool generates 4 times the code with the same *percentage* of errors, then there are four times as many errors also. Even though the error rate is unchanged as a percentage of the total code, and the cost per error generated declines due to the new productivity factor, *the total number of errors generated may increase four times if the organization continues to employ the same number of programmers, engineers or designers.*

This is almost counter intuitive, but not quite. After a moment of thought it becomes clear. If the organization used the tool to reduce the labor (cost) going into feature development (lines of code in the example) by a factor of four, then the productivity tool would reduce costs and costs per error, but generate more errors which ultimately have to be resolved in some fashion.

In that case the impact on crash rate from the C_d term is to increase it, though the actual mechanism may be circuitous, through increasing the number of features and therefore V_f. If costs are not reduced, but more features are added, these invariably contain additional errors, and the very smart productivity tool or method has increased the crash rate.

Cost per feature has declined. Oddly, we find that cost per feature is not a parameter to the model. The value of the features V_f is a parameter. But if no new features are added and the features cost less, the company will eventually be obliged to lower the price (presumably, through competition), and so V_f may decline. As the market volume increases due to the lower price (supply-demand effects), the incremental value of the features becomes less in the market, and the crash rate will go down. We will learn below this is one of two reasons why commodities are usually reliable.

The defect ratio D directly measures the effectiveness of the testing process. I don't say "efficiency" because no costs are considered. Both development and testing costs are wrapped up in C_d. If new productivity tools are applied to finding and correcting errors, the results are less ambiguous, but there is still room for a few surprises.

If more errors are found without increasing costs, then D will become a small number (it is usually 1 or less), in the denominator of a denominator term, suppressing crash rate.

If more money is spent on testing, with or without finding more errors, crash rate declines, but less dramatically since there is no leverage involved as there was with D.

If testing productivity increases, one finds the same number of errors or defects as before, while spending less money, then crash rate actually increases, not dramatically, but it does, since C_d declines.

Thus the surprising effect is that productivity is like any very high powered tool. You have to be very careful how you use it, or you will cut down the wrong thing and crash rate will increase.

The relation to Bayesian updating

The kind of probability most people are familiar with, used in medical studies for example, is Gaussian probability together with simple correlation. It has no knowledge of history, and requires a great many instances to make inferences. So for high value failures that occur infrequently it is of no direct use. A large part of the field of risk analysis is an attempt to calculate the probability of high value failures from traditional statistical data on individual components. But this does not account for all the "vague" interactions that contribute to the high value failure (discussed previously), nor are the statistical data on components for new technologies operating in new environments always available. Thus some means are sought to improve the ability to learn from smaller sample sets.

One of those methods is Bayesian probability, in which one starts with an estimate and updates that estimate with each new event. We assume the reader is either familiar with Bayesian probability, or will investigate it separately. We only discuss here qualitatively the similarities and differences with crash rate theory.

Bayesian updating is a statistical process, except for the initial estimate. There are no economic considerations in the updating. There is no relation between events of different magnitudes in Bayesian updating. Most textbook examples are binary, i.e. criteria are adopted to decide whether an event, such as a crash, occurs or not. If foam falling off an orbiter and striking a wing occurs on every mission but does not cause a "crash," then Bayesian updating gives a misleading feeling that the vehicle is getting safer and safer. What is actually going on is that the Bayesian updating function is just "averaging" the events it detects, and when the big crash finally comes, it will adjust to the right answer. Too late.

In describing how crash rate theory is different, we also must admit that it requires the intelligent application of heuristics. It is not an autonomous mathematical formula that analyzes data for the project manager. The project manager has to decide what data to collect and examine. However, crash rate theory provides a clear method for using a variety of data that Bayesian methodology does not. While it might be possible to modify Bayesian methodology, one would have to add the connection between failures of different sizes, and economic factors, and we speculate that something like the crash rate model would be the end result.

The defect ratio can be calculated on defects of any size, or on all defects. There is still the question of what qualifies as a defect, but it is easy to be on the safe side and classify relatively minor items as defects. Changing the classification will not necessarily change the *ratio* between how many are found during testing and after deployment, unless the testing/inspection process is overly specialized for a few types of defects. A separate independent testing process is a handy gateway across which to measure this value, otherwise the test instrumentation has to be maintained into the operational life of the system, which I recommend. In the case of manufactured goods there is usually a continuous inspection process, and later we will discuss the relation between crash rate theory and continuous improvement, sometimes called Six Sigma.

In summary, crash rate theory obtains crash rate as an economic constraint, not as a statistical process, and provides therefore additional insight that can be used to relate small and large defects, and provide information about low rate large scale failures that is difficult to determine in advance using statistical methods. At least, that is what we suppose. Now we turn to our survey of product categories and application areas to see if crash rate theory makes sense.

Commodities

For commodities, with certain exceptions, production and marketing costs M approach the market value V_f. This is the essence of a commodity. It is both universally needed, and able to be produced without esoteric technology or risks. Reports of people dying from consumption of bad rice or wheat are extremely rare. Sugar and salt have health effects when consumed in excess, but producers are not held liable so those costs do not enter the crash rate model. Asbestos ceased to be a commodity after producers were held liable. In general, the small numerator implied by $M{\to}V_f$ implies a low crash rate.

The cost per crash C_c for a commodity is often enormous. This also keeps crash rate low, since the term occurs in the denominator. A good example is the energy industry. Energy (e.g. oil) can produce massively damaging spills and violent community-impacting explosions. Crashes of the magnitude of the BP Gulf spill or earthquakes from fracking will act as a risk deterrent. Imagine if producers were held responsible for climate change! We do not yet know if the world will succumb to global warming, but even if it does the crash *rate* will be extremely low, i.e. $R_o \approx 1/ever$ crash in all of history. Such a low rate may violate the statistical equilibrium assumed by our derivation, but qualitatively we can understand its effect.

Technology and software

At the current time, 1,000,000 transistors cost about as much to make as a single grain of rice. Software which is widely distributed on the internet and developed by open source organizations is essentially free, and most crashes are merely inconvenient. So M, C_d and C_c are reduced. This suggests the crash rate for software and technology will be enormous, and it is. Who has not dealt with dropped calls and a computer that has to be constantly re-booted?

For software that must handle financial or life critical functions, costs are enormously greater. Software productivity for manned spacecraft control can be as low as one source line of code (SLOC) per day, fully verified. Productivity can be hundreds of times higher for non-critical applications. So the crash rate model is quite consistent with our general shared social experience with technology.

However, in the denominator is one surprise, the ratio C_d/D. The surprise is not in the ratio itself, but in our use of it. If the cost per defect during development C_d increases, the denominator increases and operational crash rate decreases. In other words, testing expenditures can decrease crash rate, because they increase the denominator without adding any more features (V_f). If the testing process is very thorough, resulting in a low defect rate D, the term C_d/D increases and again crash rate is lowered.

However, according to the model, if a company engages in "process improvement" and invests in new development technology that reduces the cost C_d by economically finding and fixing bugs during development, and other factors are constant, then operational crash rate could actually increase. To remedy this problem, the company would have to test to a lower defect ratio than previously, thus doing more testing, which still might cost less. But to lower engineering costs and test only to the same level as before should lead to a higher crash rate.

For example, many people wonder how a large and very capable software company can produce version after version of its flagship product, claiming each time it has fixed stability problems. And yet users find stability problems remain and in some cases increase. Software development has a long history of improvement in the process of creating and debugging software. Hand coding gave way to assemblers, then compilers, on-line editing and debugging, structured programming, object oriented programming (OOP), application programming interfaces (APIs), and so forth. Each advance in software or hardware led to an array of new features which increased usage of software systems to the point where now most people have several computers, and use their phone as a computer. What used to require a command line interface with a precise protocol is now done by vague flicks of the finger or thumb across a touch screen. The software must handle and interpret input from any screen object at any time, not just the next valid command. Even pocket devices are expected to do many things at once, some of which involve real time processing demands, vast storage requirements, and interaction with many other computer and network systems developed at different times by different vendors using ever changing interface standards. It's not surprising in that context that more testing is required.

Extension to the public sector

Our analysis has been microeconomic based on a profit relation. However, we have avoided specifying how the decision is made. It could be made by market forces, with more profitable organizations gradually gaining market and capitalization share in a manner similar to Lo's [2013] adaptive market hypothesis.[40] Or it could be the result of savvy or lucky decision making.

The results can be extended to non-profit and governmental organizations in several ways.

First, there are explicit pressures to allocate funds efficiently between governmental and private organizations. These decisions may not always be rational. They are essentially political. But at least there is consideration and debate of whether funds are best expended through the public or private sector.   This puts pressure on public sector managers to reach an efficiency level comparable to the private sector. They may not perfectly arrive there, but there is pressure.

Second, the public and private sector draw from the same pool of employees, vendors, and management consultants. They tend to do things in ways more similar than critics of either sector realize, even if there is no particular reason to be similar. So the crash rate model can be extended to the public sector by cultural similarity.

Third, the public and private sector draw on the same technological pool for methods and processes of doing things. They use the same kinds of computers and software, vehicles, and so forth. So parts of their development and testing methodology and costs are very similar.

And finally, even government agencies may in the long run have to absorb costs of their mistakes. In the 1980s after the Space Shuttle Challenger exploded during ascent, funds were appropriated for a replacement orbiter. But the NASA budget was not fully increased by that amount. The development of a space station and other programs were delayed. Following a second catastrophe in the 2000s, the loss of the Columbia, the flight rate was reduced, limited to safe haven destinations (mostly the International Space Station), and by the end of the decade the program had been canceled, at least in part because of safety concerns. The function of supplies and crew transportation to the space station is now to be provided by the private sector (sort of ... the government still writes requirements, lets contracts and monitors and administers them). Both safety and costs are components of this discussion, which indicate the factors represented in the crash rate model are certainly influencing this large government program.

Role of culture and training

The defensive driver training courses mentioned earlier are based on increasing drivers' awareness of the true costs and probabilities they face, and are backed up by reams of data and examples from real life. A safety culture online class I encountered recently, which might be similar to one used in your company or institution, provides both a useful definition of organizational culture, and a contrast with the methods and objectives of driver training.

Culture is simply *"the way things are usually done."*   This reflects assumptions about the economic costs, and driver training works by improving accuracy of those assumptions. The culture training made statements about the safety culture that did not seem to agree with the experiences of mature employees whom I interviewed, and none of the examples were drawn from real life. Probably there was a "hope" of making some change in the culture. In our model, high cost training could change crash rate, but low cost training could only improve knowledge of it.

Employees will notice if budgets are allocated and careers made by bringing up reliability issues, or not, and will act accordingly.

Relation to risk compensation

The crash rate model differs from risk compensation in several important ways. We already discussed that the crash rate model is corporate, and risk compensation is primarily a theory of individual action. They can both yield an unexpected rise in incidents (crashes, failures or accidents), but for different reasons which we will compare.

In the crash rate model, crash rate can unexpectedly rise when development processes improve, and the cost of finding and fixing problems decreases. To put this more strongly, when the cost of adding new features decreases, corporations will chase larger profits by adding more new features. This will cease when the impact of operational failures offsets the value of the new features. The decrease of required effort for some activity (new development) leads to more of that activity. The increased activity (functionality, features, number of customers, and usage by customers) leads to the increased crash rate.

In risk compensation, the decrease in risk (rather than cost) leads to similar increases in activity, which leads to an increased crash rate. The increase may take the form of, for example, driving more miles, in which case the crash rate per mile might not increase but the total crashes would increase. Or it may take the form of faster activity, faster driving, and more efficient use of time, which can lead to crash rate increases by all measures.

So the crash rate model is driven by cost, and risk compensation is driven by risk. The crash rate model is corporate and risk compensation is personal. But Corporations are composed of people. And risk can be related to costs using the expected value principle. Once the risk/cost equivalence is made, from there the models operate in a similar way to produce the unexpected results.

It is possible to go even further in comparing the crash rate model to risk compensation, by noting similar roles for technology.

Not always, but often, safety improvements are the result of technology. Certainly air bags are technology. Fire safety blankets are through time improving with technology. Global Positioning System navigation devices and cellular phones are dramatic technology devices that can be used to improve safety. Do people take more risks because of these devices? That is the premise of risk compensation. Do corporations make more of these devices because people will buy them? That is the premise of capitalism and profit seeking. Do corporations use these same devices and technologies to improve their operation and develop even more features and devices at lower cost? Yes they do.

Risk homeostasis is only partly comparable to a microeconomic model, with clearly psychological elements. However, that does not mean crash rate is all economics and no psychology. We have discussed how, indeed, it can be applied to the public sector partly on psychological (cultural) grounds.

Air travel safety

As noted by Carlsson et. al. [2004], we are willing to pay more for safety in the air, or put in terms of our model, the crash rate for air travel is lower than expected, based on its cost and value in comparison with other types of travel. There could of course be psychological reasons for this. People may fear the lack of control over even mundane failures, and may choose instead to drive their own vehicles. In the survey conducted by Carlsson, this is what travelers *said*. However, the survey also indicated that people are willing to pay more for a given safety improvement (from 1.0 in a million to 0.5 in a million chances of serious mishap) when the cost of the trip was higher. The comparison was between air and ground taxi, not a private vehicle in which travelers might be willing to invest more.

We do know that air travel is very safe compared to automobile travel. Can we identify reasons consistent with our model? There is a difference is the cost of a life in an airplane and an automobile crash. Most crashes result in lawsuits only against one of the parties, and with minimum insurance, the recovery may be only $30,000, vs. $2 million for an air crash victim, a 70 times greater C_c. Assuming an average of $60k for half the deaths, the ratio is still 70x. It is little surprise, then, that lifetime odds of dying in a car are 70 times greater than flying.[41]

It is possible that the use of air travel violates the assumption of competitive economic pressure on the user (traveler). Slightly more than 50% of all air travel was for pleasure or leisure in 2001, according to the U.S. Department of Transportation [2008].[42]　This kind of recreational travel is different than, for example, racing or paragliding. It is not the travel itself that is recreational. The travel is only a means to an end. Possibly the economic utility of recreational travel is not as great as business travel, and users will avoid trips which appear to be dangerous. Even in regard to business travel, the age of internet has made competing options such as video conferencing very cheap, easily available, and generally accepted by business partners. So before applying the crash rate model, one needs to determine the validity of the assumptions, and the true field of competition.

Strictly within our model, we see that the cost of testing airplanes is very high. The cost of crashes, and even relatively minor (non-fatal) incidents if they occur in multiples, can be extraordinary. For example, at this writing the fleet of Boeing 787's has recently been grounded for an extended period because of failure of a lithium ion battery module to contain failures. Deliveries of 787's were on hold and Boeing expended additional funds for testing. Meanwhile customer airlines could generate no revenue from those aircraft.

The cost of multiple crashes is likewise high. After 4 crashes, the de Haviland Comet 1, the first commercial passenger jet, was taken out of service from 1954 until 1958, when the third revision (Comet 4) was introduced. 1959 was the last year a new jet product was introduced with the brand DH, though the company was acquired by Hawker Siddeley which developed a successful business jet. It was sold again and these are now manufactured in the United States. Many of today's air travelers, however, will still remember and have a negative association to the airline brand ValuJet, which began in 1993 by buying older aircraft, and after many safety incidents and harsh criticism from the FAA had a notorious fatal crash (Flight 592). The airline then had serious financial problems, and merged with AirTran, taking the name of the smaller carrier. So the cost of a crash can in some cases go far beyond the equipment lost and the compensation to injured or deceased customers.

It is interesting to note in this context that if ticket costs are taken to be representative of V_f, low ticket costs would tend to push crash rate down. The world's largest low cost airline, Southwest Airlines, ranks 21 out of 60 on the JADEC [2012] safety rankings.[43] But to the extent low cost carriers hold down recurring costs M then this effect is reduced as value to the company of each flight remains high. The real leverage is in the defect ratio, and whether an airline performs inspection and maintenance is critical. Note that Southwest recently acquired AirTran. Safety ranking information for AirTran was not readily available at this writing. Five of the seven largest low cost airlines rank in the top half of the JADEC rankings, ranging from 14 to 23. One airline GOL spoils the overall ranking of the group. Nevertheless, the data are sufficient to conclude that low cost does not have to mean unreliable, and we at least keep open the idea that V_f may indeed play the role we suggest.

Comparison to "Six Sigma"

The origin and nature of the term "six sigma" is almost lost, along with the memory of the dramatic effect it had on the transfer of manufacturing from American to Japanese companies, and the global transfer of wealth to Japan. Popular Six Sigma associations and websites (Six Sigma Online[44], iSixSigma[45], Process Quality Associates[46]) give histories that begin with Motorola in the late 1980s. This was actually *after* the dramatic transfer of industrial might to Japan, especially in the auto industry, that prompted President Reagan to forcefully negotiate a revaluation of currency exchange rates with Japan, and his anti-regulation administration passed sufficient tariffs to force most Japanese auto manufacturers to open plants in the United States. While this was going on, American auto manufactures were still engaged in "planned obsolescence."

1979 Honda Civic   1979 Chevy Nova

Detroit was unimpressed by the 1979 invaders[47]

Only in one of the three histories above do we learn from a brief anecdote about Japan that in 1970 they acquired a Motorola television factory and decreased the defect rate to 5% of the previous value. Motorola brags about taking 17 years to learn from this? The author owned 3 Motorola RAZR phones in a 14 month period, all of which broke, not counting one which was delivered broken. The lesson forgotten, Motorola Mobility was sold to Google and then to the Chinese company Lenovo. No one could fix the culture of features over quality, and in retrospect the matchup with feature-innovator Google was doomed from the start on this account. Google's Android phones were not considered as feature rich as iPhones, but were produced by quality-focused Asian companies HTC and Samsung. The author has owned a steel-cased HTC phone for more than 3 years, dropped it several times and even left it in his pocket to be laundered in the wash, and it still works fine.

The Japanese were heavily influenced following WWII by the American engineer Deming, who did not exclusively take the dry statistical approach that is attributed to Six Sigma in the post-1987 American academic and industrial literature. He didn't even use the term six sigma. Instead he emphasized philosophical viewpoints represented by highly abstract equations (Akpose[48]) not unlike those of our model, such as:

Quality = (Results of work efforts) / (Total costs)

When people focus on quality, Deming said, quality tends to increase and costs fall over time. When people focus on costs, costs increase and quality declines over time. The author has witnessed 40 years of the steady creep of "full cost accounting" in both government and industry over his career, with exactly the results that Deming predicted [the author's qualitative impression].

It is counter-intuitive that costs should fall when one focuses on quality. Surely there are some limits to such a notion? While this is well outside our current scope, if we compare Deming's assertion to the prediction of our model that spending money to test to a lower defect rate (higher quality) highly leverages the funds expended and lowers failure or "crash" costs, the prediction does not seem so farfetched.

Application to other examples

The **Montana and Autobahn motorway death rates** discussed earlier are due to individual, not corporate, choices, and require still some psychological elements in their explanation, either risk compensation or risk homeostasis or both. The quantitative input from the crash rate model is from the changing ratio of the value of the speed (which we take as a time value, assuming motorists are not racing or thrill riding routinely) to the cost of crashes, V_f / C_c. The value of speed rises slightly while the cost of crashes in property, health and liability terms rises dramatically. This explains a reduced crash rate, but not necessarily the reduced fatality rate. The factors already mentioned such as attentiveness, which are the likely reasons, are psychological factors.

As far as government response to these effects, we find that cost numbers are explicitly assigned to loss of life and compared with costs of mitigating the loss of life (possibly except in Sweden), which can't be taken as confirmation of the crash rate model only because we used this example to derive assumptions for the model. So provided our math was correct, agreement with the corporate-government response is almost a given.

For **no-fault auto insurance**, the cost to the at-fault driver's insurance company, and therefore to the at-fault driver in terms of premium increases, is probably about half of the total accident costs. The only additional assumption we require is that a large fraction of accidents are caused by a few drivers. We then have exactly the right factor in cost (half) to explain the approximate doubling of premiums. A quantitative confirmation of the crash rate model is obtained.

For **seat belt and airbag effects**, we find that engineers and accountants calculating fatality reductions did not consider any change in user behavior. In this case, psychological factors are not needed, only that the users are economic optimizers as are corporations. Given a lower cost of crashes, they increased their speed or other risk taking which was economically useful to them, and approximately restored the old balance, but not necessarily the old fatality rate. Therefore it appears as risk compensation, but it has an economic explanation.

In addition to the user optimization effects, with regard to airbags it is likely that the original analysis was not done correctly to account for poor positioning of unbuckled occupants during crashes, and similar effects, and a revised analysis after a fact finding period would probably be more accurate. Airbags are of low value if occupants are not buckled up, and most of the protection comes from the belts.

There is also a transient period of adjustment as users (drivers) realize airbags do not protect them from side or angle impacts, and either adapt their driving precautions or acquire cars with side airbags, both of which are occurring.

Analysis of the **war on drugs** is quite interesting because of unsuspected effects of other policy changes.

- User crash costs C_c were made lower by medical advances and the late 1980s ruling that emergency rooms could not deny treatment.

- "Just say no" commercials tried to convey a complex message, that invitations to use drugs from "cool" people should be turned down. The overlooked but strong subliminal message, which is especially appealing to teenagers, is that "cool people are doing drugs."

- Incarceration costs to drug dealers went down from two unrelated effects. Snitch laws enacted to catch drug dealers worked in reverse, as the dealers had a larger pool of names to snitch on than users. In some cases it is reported that users helped entrap friends who were not actual users. Costs borne by non-users or rare users do not even appear in the crash rate model and have no effect (except possibly on the choice of friends). And finally, reduced penalties for minors induced drug distributors to recruit minors as dealers.

- Supply-focused enforcement costs are borne only by society at large, and have little effect on the drug supply chain. The costs to make it necessary for smugglers to use high-value throw away vehicles, for example, and even submarines in some cases, are even higher. An entire border has to be watched closely. But the smuggler only has to get through in one place.

The use of drug testing is interesting because it should take advantage of the leverage of the defect ratio to multiply the effectiveness of testing costs. For this to work, testing costs cannot be trivial, and the tests must be accurate enough to actually obtain a good defect ratio (i.e. few drug users escape notice). If there are few drug users to begin with, the testing program will be expensive. As with other applications of crash rate theory, good data are hard to get and if one had good data one could fix the problem by simply disqualifying the drug users. In other settings testing has been limited because of the risk to our personal liberty.

In the case of hard drugs, there is a significant probability of death from overdose, or loss of employment and wealth from overuse, or death from conflict among dealers. Legal penalties barely increase such costs, so it is unsurprising that brute force cost has not worked.

In the case of **fire safety blankets**, it appears that like motorway deaths, this is a case of psychological factors, mostly an incorrect perception of the cost of entrapment, and not directly analyzable by the crash rate model. However, as before, government response such as discontinued use of the blankets in British Columbia in 2005, reflects a rational economic analysis as we have supposed.

The topic of **nuclear energy** is interesting. Approximately 1.8 million lives have already been saved by this technology.[49] [50] [51] [52] Consider the data in the table below.

type of energy	fatalities	% of world electricity
Nuclear	5163 total	12.3% from 437 plants
Fossil fuels	300,000 per year	69.4%

Energy related fatalities including radiation leakage & pollution

The data include Chernobyl and projections from Fukushima. They do not include acts of war, which would be quite large from both sources. The world is still "learning" about the crash rate and impact of nuclear energy. Since the crash rate is rather low, the learning rate is slow and we are nowhere near a static equilibrium. The perceptions of nuclear energy as a terrible disaster may come to pass, or not, but presently they are just perceptions. The author will predict that if the disaster rate does not get worse than it has been, nuclear energy will grow in use dramatically. Even Japan is returning to nuclear power only three years after Fukushima.[53]

The author attempted to do some analysis of the **Shuttle orbiter** in 2003. V_f is not directly measureable for a non-profit space project, but we can probably infer it from total program costs. C_d and D are not known from easily accessible public records, or even from internal NASA records. Large testing costs were incurred for engines, avionics, and tiles. Poor defect ratios were corrected and no loss of life was ever attributed to a primary failure in these systems.

The solid rocket boosters (SRBs) and insulating foam were considered mature technologies, and many operational defects were ignored (cracked and burned out seals, falling chunks of foam). Both produced fatalities.   Consideration of these findings prompts recommendations for how to use the crash rate model in development and operational processes, which we will cover in the next chapter.

After 1986, military and commercial users withdrew or were dropped. This possibly changed R_o from 1/50 (based on the Challenger crash) to 1/84 (84 more flights until the Columbia crash). While not statistically significant, it is consistent with the crash rate model. If the value of the features goes down, so does the crash rate.

In 1986 there was a change from a quasi-military crew to civilian (a teacher in space), without re-examination of R_o. As it turned out (of course this was not known beforehand) the crash rate of 1/50 is comparable to military but not civilian endeavors. WWII bombing runs over Germany had a crash rate of approximately 1/25, and this was considered very hazardous duty.

Between planning in the mid-1970s and commencement of operations in the mid-1980s, there was a 100 to 1 reduction in flight rate. How would this affect the crash rate R_o which is expressed as "crashes per some number of flights"? We will analyze this in the next chapter.

As we suggested earlier, crash rate theory cannot solve a knowledge problem such as the **war on cancer**. There may be other aspects of the problem it can shed light on. For example, testing was (logically and in accordance with the theory) relied upon to screen for defects, to identify cancer early when its treatment was more likely to succeed. The U.S. Preventive Services Task Force concluded that *"false-positive test results, overdiagnosis, and unnecessary earlier treatment are problems for all age groups, false-positive results are more common for women aged 40 to 49 years, whereas overdiagnosis is a greater concern for women in the older age groups."*[54] Not only are the tests not good enough to improve the defect ratio, they are inaccurate enough to make it worse, and the treatments are sufficiently harmful that it is not advisable to treat false-positives. So what we have is $D \approx 1$. Many or most defects become operational defects. The leverage term is ineffective.

Defect screening may have played a role in the crash of Germanwings flight 9525 in March of 2015. The pilot had left the cabin. The cabin door is impenetrable to screen out "terrorist" defects. But the co-pilot was unresponsive in audio recordings, and the pilot was "screened out." After an 8 minute dive 150 people were killed. Media articles have so far focused on whether the airline's psychological defect screening was adequate – a process known to have poor accuracy – without noticing that previous defect screening measures kept the pilot from intervening. It turns out the co-pilot had researched the door security.

Applying Crash Rate Theory

Options for influencing crash rate

The crash rate model suggests that companies competing with each other in a market, using similar technology, must converge on similar crash rates. It is possible a company can make a marketing issue out of the reliability of its products, and this has been done for automobiles, phones, computers, and many other devices. In that case V_f increases if R_o decreases, an external condition which opposes the model's natural equilibrium. It appears that a constant external pressure would be needed to *keep* the model off equilibrium.

If an executive at the top seeks to influence a large organization to have a specific R_o, how can it be done? The crash rate model suggests improving development processes runs some risk of backfiring, if middle managers use the advantage to seize market opportunities, and add too many features. However, spending money on testing and reliability, that is, on making a very low defect ratio D, seems to be an unmitigated benefit. It inflates the C_d/D term in the denominator, reducing the crash rate. Anytime cost cutting measures are applied to M, a corresponding effort to increase C_d/D, or alternately passing the savings to the consumer as lower V_f, would be needed to prevent the crash rate from rising.

Indeed, testing and quality control, along with low prices, in general work, and not just at reducing crash rate. For several decades, this seemed to be the strategy of the entire nation of Japan. Testing and quality control (though not low costs) were also a premise of the early space program. However, in the last two decades of globalization, all costs have been under attack, and costs ranging from testing to research have been deemed ancillary and have been cut. The result is reductions in M and C_d which, according to our model, will increase crash rate.

A Project Manager Exercise

In this section we consider a hypothetical example, but one based on reasonable extensions of existing systems in ways that are currently being planned to one degree or another. The author selects an example in the aerospace field where he has 42 years of experience with which to judge the reasonableness of the scenario and projected results.

The project is a commercial transport to an orbital hotel/tourism facility. Some background will establish the degree to which the project is realistic and within a realistic infrastructure environment.

Bigelow Aerospace[55] has acquired the inflatable TransHab technology from NASA, hired NASA personnel, and announced plans to build an orbiting hotel. The TransHab concept was developed at the Johnson Space Center in 1997, and a full scale prototype was built. You can read a description on NASA's history website.[56] In micrometeoroid tests the several inches thick shell with woven Kevlar was superior to aluminum shell modules because of the greater stopping length. It was intended for a Mars mission, but of course would have been tested on the International Space Station (ISS). Boeing became concerned about the future of an aluminum module they were building, and lobbyists persuaded Congress to write into NASA funding legislation that no funds were to be spent on inflatables. Rather than throw away the technology, it was given to Bigelow.

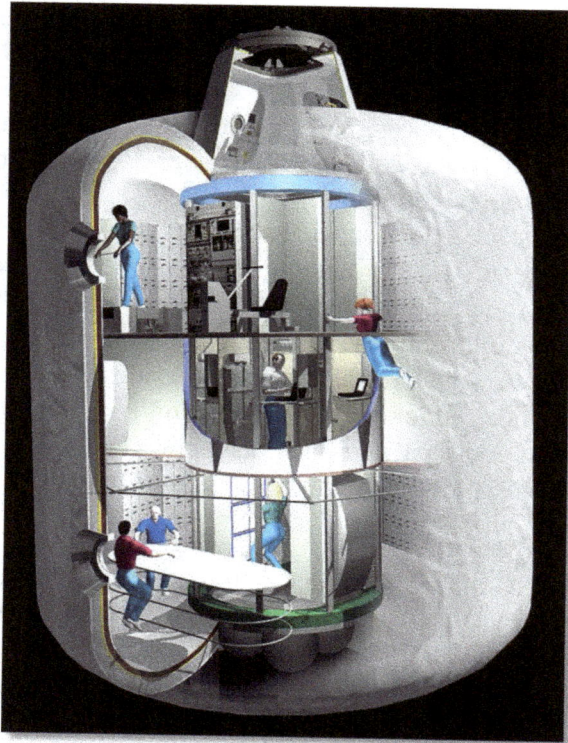

TransHab cutaway[57]

Artists concept of TransHab attached to ISS (NASA)

The TransHab could be launched in the Shuttle cargo bay without need of a heavy lift launch vehicle. Interior outfitting would be launched on a subsequent Shuttle flight. The author worked on the project briefly in an avionics instrumentation capacity. It is not unusual for NASA engineers now retiring to cite it as the best project of their entire careers. It looks like NASA has, 15 years later, decided to pay Bigelow $17.8 million dollars to further develop the technology,[58] which is normal procedure for developing government technology.

A few years later the author developed a mission concept called Tele-Exploration using the TransHab, the VASMIR engine, tele-operation, and Robonaut technology for a $12 billion Mars orbital manned mission, and from this obtained substantial experience in estimating the cost and schedule of complex, integrated space systems. The author also has had training in space architecture from the University of Houston, in addition to his regular work experience which included managerial responsibility for major avionics integration and test facilities. I hope to write a public description of the Mars idea in the near future, but it is off-topic for the present book. However the experience gained is relevant to constructing a realistic example.

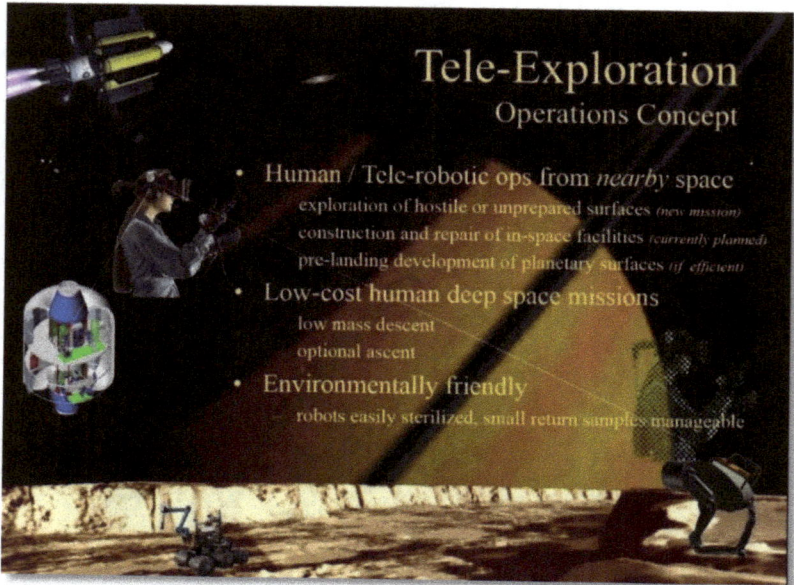

Tele-Exploration concept summary

Current NASA plans for access to the ISS are not fully private and not the basis of the example, but for reference I'll mention that these plans are called the Commercial Crew Program (CCP). Currently two bidders have been selected for further development, SpaceX and Boeing. Both designs are capsules and neither will take private citizens into space.

Instead, the example is based loosely on an extension of the idea of Virgin Galactic's (VG) Space Ship Two,[59] Sierra Nevada's Dream Chaser, and similar projects, of which there are many. Virgin Galactic is the furthest along of the purely private projects. It is a partnership between Richard Branson, Microsoft's Paul Allen, and Scaled Composites' Burt Rutan. It has a crew of two and is intended to carry 6 passengers to an altitude of about 70 miles at 2600 mph. At that speed it will not need a sophisticated heat shield, and requires only a tiny fraction of the energy required to achieve orbital velocities.

The initial ticket price was $200,000. The development cost estimate in 2007 was $108 million. At that time VG had 200 paid customers ($20 million dollars). But by 2011 the projected development costs had risen to $400 million, rivaling the overruns of any government project. The ticket price has now risen to $250,000.

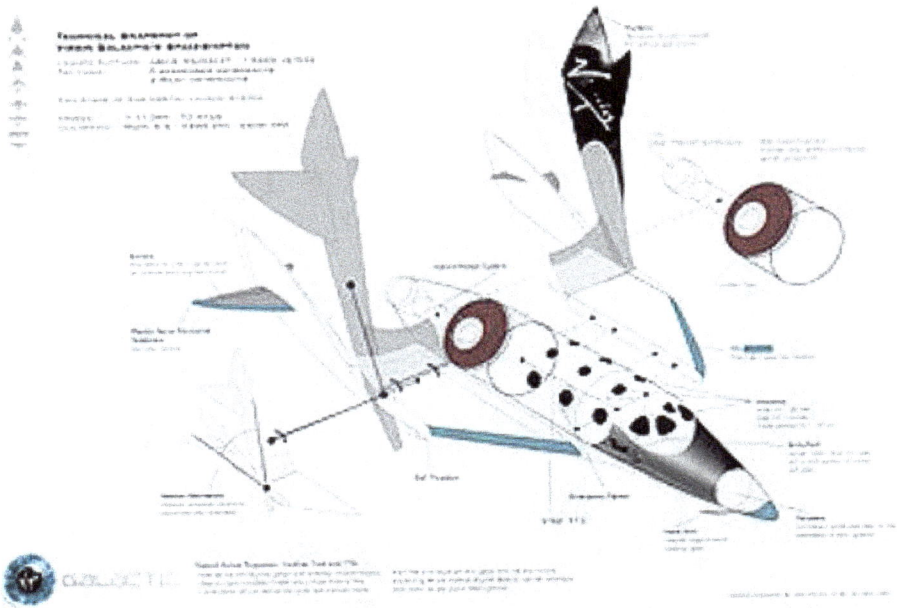

Space Ship Two technical diagram[60]

The development costs for the exercise were scaled from VG's original estimate to a more capable vehicle. I have not tried to re-work the example with revisions. Probably some lessons have been learned that would reduce costs on a follow-on vehicle, and if the reader disagrees, which would be perfectly reasonable to do, it is easy enough to scale the numbers as you read the example.

Also since constructing this example, the Space Ship Two project had a mishap in which a vehicle was destroyed and one pilot killed. The investigation was not complete at this writing. But the scenario that played out in real life is eerily similar to the hypothetical one that was envisioned.

A second relevant program is Sierra Nevada's Dream Chaser. This vehicle will go all the way to orbit and is projected to have development costs of less than $1 billion according to a recent space.com article, [61] only about twice VG's now-projected development costs even before considering crash recovery. The same article gives amounts for awards for only partial completion of Boeing and SpaceX capsules of $4.2 billion and $2.6 billion respectively. Sierra Nevada's costs are based on design and manufacture by *"more than 30 states by over 15 strategic partners and over 30 other subcontractors and suppliers"* according to Sierra Nevada's website,[62] which seems to suggest there might be room to lower costs with a more consolidated approach. Unlike a capsule, the craft can land on any commercial runway in the world. The company says it has plans to continue development even without NASA funding. [Ibid. 62] The Dream Chaser, like capsule designs, is based on NASA and military technology such as the X-38.[63]

Dream Chaser[64]

The goal of our hypothetical Space Tourism Transport (STT) is to visit an orbital hotel, presumably a future version of Bigelow's efforts. Tickets will be $1 million, or 5 times the original price of VG's suborbital tickets. Market surveys indicate 5000 or 1% of the 640,000 people in U.S. in households of over $20 million net worth will be interested over the next 10 years, and many more outside the U.S. So for individual passengers $V_F=\$1$ *million*. The objective is to hold recurring costs (*M*) to $750,000, giving a net revenue of $250k per passenger, times 10 passengers, or $2.5 million operating profit per flight. In 2015 launch costs are about 6 times higher than this for an 11,000 kg payload.[65] Our exercise is set a few years in the future when a reusable launch vehicle (RLV) is available with 10x lower launch costs and 400x improved reliability.[66] Our analysis does not include the RLV or the stay in orbit.

Two vehicles will be built at $250 million each. One flight a week gives an expected operating profit of $125 million per year. Investors expect to go public at the end of the first year of operations at a price to earnings (PE) multiple of 40, giving a capitalization value of $5 billion dollars, a ten-fold return on their $500 million investment, typical of venture capital expectations.

Passengers sign a waiver of liability, but it is not expected to hold up in case of vehicle systems failure. At-fault accident liability is estimated at $120 million dollars (~$10 million per occupant including pilots), which no one will insure at a reasonable price. Investors insist on an $R_o < 1/(25\ years)$ to insure public confidence and some profit after recovery of development and liability costs. Five year operating net revenue yields $625 million, vs. development costs of $500 million, and $120 million liability once in 25 years in the worst case of $R_o = 1/25yr$.

The reliability goal of less than one crash in 25 years, or R_o $<= 1/1250\ flights$, is still 1000 times more risky than a 1000 mile automobile trip, but it is 50 times safer than a WWII bombing run over Germany and 25 times safer than riding on a NASA Shuttle.

The dilemma to be solved

Test flights are $10 million, slightly more than the $7.5 million of operational flights because of instrumentation and engineering costs, and general startup inefficiency. After 10 test flights the team finds 5 major but ultimately fixable problems. This gives $C_d = \$500mil / 5 = \$100\ million\ per\ major\ defect$. You are the project manager. You have to do some guessing at this point. By that I mean you have all your engineers conduct a traditional risk analysis, using all the data you have found, and you conclude that most likely $D <= .1$, that is, you think in the next 50 flights (one year of operation) only two and a half serious problems are likely to show up. (D of .1 times 5 defects per ten flights times 50/10 = 2.5) In crash rate theory risk analysis is still essential, but assigned the more circumscribed role of forecasting the defect ratio.

We now have all the parameters necessary to calculate R_o, so we plug in to the crash rate equation:

$$R_O \approx \frac{V_F - M}{C_C + C_d / D}$$

$$R_O \approx \frac{\$1M - .75}{\$1B + \$100M\ /.1} = \frac{.25M}{2B} = \frac{1}{8000}$$

At first glance you are quite pleased with that number. But immediately two things happen:

- After you enthusiastically reveal the calculation at a press conference, a space.com reporter who is on the ball realizes that you have used the V_f for a single passenger, so you really expect a crash in every 800 flights, not 8000.
- A competing spacecraft fails spectacularly, driving up the perception of risk, and ticket sales top out at 100.

The crash rate is not bad for a spacecraft, about once in 16 years, but worse than the original requirement of 1/25. What if the crash comes sooner rather than later? And of course the real problem is public perception. Ticket sales barely account for two months of operation and no or few new tickets are being sold.

What do you do?

The alternatives

A hedge fund full of bright analysts, probably former physicists (as sometimes happens with real hedge funds) who understand numbers but not human nature, and who are enthusiastic about space and "willing to take risks," offer to buy the company. Their plan is to charge $4 million per ticket and conduct only 10 flights per year. That recoups development cost of $500 million for the two vehicles in 1.5 years (the revenue is mostly profit at that ticket price). This is one alternative. Here is the analysis of the expected crash rate:

$$R_O \approx \frac{\$5M - .75}{\$1B + \$100M\ /.1} = \frac{4.25M}{2B} = \frac{1}{470}$$

Due to the low flight rate and high financial pressure on each flight from the five times higher V_f, the new plan expects a crash about every 4.7 years (one crash in 47 vehicle flights, one in 470 passenger flights).

Do you take the offer? Is that the best you can do? Is it even making a change in the right direction? Your investors are getting nervous, and seeing a cash offer from the hedge fund are ready to bail. But they are also space enthusiasts and have deep pockets. Everyone realizes this is a full scale retreat, merely a plan to get out of Dodge City before the gunfight breaks out.

You try to figure a way to make the vehicle safer with improvements to the systems that had defects, and more testing, including:

- $50 million in hardened space avionics, thermal protection upgrades, on-orbit inspection and repair capabilities including a spacesuit, and instrumentation for continuous monitoring of operational flights
- 5 additional test flights for $50 million to build confidence and verify defect ratio

The additional investments will add $100 million to C_d. You think the defect ratio will come down, but not in areas where there is no design margin, so you only assume a 20% improvement to $D=.08$. The price is only a 20% cost increase for your original investors. If it gets ticket sales going, they will probably buy it. Increasing the number of test flights will impress the public. Each flight will be highly visible. A quick poll taken by a radio station personality suggests ticket sales will resume. The new crash rate calculation is

$$R_o \approx \frac{\$1M - .75}{\$1B + \$120M / .08} = \frac{.25M}{2.5B} = \frac{1}{10000}$$

This is one crash per 20 years at the original flight rate, a 25% improvement in reliability for a 20% investment increase, and probably advisable, but it is not the major jump you were hoping for. It still means there will "probably" be a crash over the projected 20 year life of the vehicle, and is still short of the original requirement of one crash per 25 years. You have picked the "low hanging fruit" technically, and remaining systems are not as easy to improve. Finally late one night you get an idea. What if I reduce the ticket price just a little bit, a paltry $100k, and combine that with an additional $250k per flight ($25k per ticket) for thorough inspection and analysis before and after each flight?

By reducing the ticket price a small amount, and adding inspectors, you reduce V_f-M by a lot, from $250k to $125k, 50%. It also reduces the apparent worth of the eventual equity offering, but with higher reliability the whole thing is less of a scam and more of a real future in space tourism, so the PE multiple might be more than 40. Let's see how this works out:

$$R_o \approx \frac{\$.875M - .75}{\$1B + \$120M / .08} = \frac{.125M}{2.5B} = \frac{1}{20000}$$

Not bad. This is the big "2x" increase you were looking for. R_o was doubled and now exceeds the original spec of one crash in 25 years, in fact it is one in 40 years, and with continuing improvements you can keep increasing it, perhaps faster than the years go by. Your investors go for it. You have secured a human future in space.

The reader probably understands intuitively the effects of spending more money on testing, and of improving the defect ratio, and of continuous improvement. But why did reduction in ticket price by itself have an effect? Where is the technical cause of this?

There isn't exactly an engineering cause. There is the assumption that no company would agree to enter this business at that profit margin if the crash rate were not less than one in 2000 vehicle flights (or 20,000 passenger flights). Just changing the ticket price might not directly affect engineering reality. But it dramatically affects operational reality, and it is possible that it could make a significant difference in crash rate. The company only loses half as much, $1.25 million instead of $2.5 million, if it outright cancels a flight. The decision to wait and avoid bad weather is twice as easy to make, and that takes off the table the cause of one of the two Shuttle disasters, and a common cause of many aircraft accidents.

The company could make further dramatic improvements by a simple technique. It could pledge $600 million of the expected $5 billion initial public offering to continuous improvement, to fixing any on-going source of defects. This would not only double one denominator term immediately, but fixing the source of defects would decrease the defect ratio further. Probably the crash rate would soar to better than once in 100 years in short order.

The company will have to be careful though. Through learning, operational costs come down. Ticket prices should come down also. Otherwise, the business will over expand and tempt fate with too many additional flights or too much pressure on each flight, and crash rate will rise again. Another temptation point will be when Lloyds decides to offer liability insurance, seemingly removing the big C_c from the denominator. In reality it is still there.

Lessons of the Defect Ratio

Before we leave this example, what do you think it would mean if several failures occurred during the 5 additional test flights? When I ask the question this way, you probably will realize that the value of D increases. As the number of failures increase and perhaps some are noticed on every flight, then the defect ratio approaches 100%. All leverage is lost, testing dollars only count at face value, the denominator becomes much smaller and the crash rate goes higher. Much higher.

Then what did it mean that foam strikes were occurring on nearly every mission prior to the Columbia disaster? When I ask it in context of the defect ratio and the fact that the disaster did happen, it suggests that the disaster was nearly certain to happen, predictable.

Recommendations for real programs

While testing is often cut from new project plans, it nearly always gets added back in after problems develop. The scenario of Space Ship Two is normal. A bunch of guys trying to build a rocket to carry 8 people for $100 million? It doesn't even sound realistic. The experience of the partners is telling:

- *Paul Allen* – developer of some of the most crash-prone software in the world, Microsoft Windows
- *Burt Rutan* – developer of one of a kind very interesting vehicles which as far as I know do not fly enough times to have good crash rate data
- *Richard Branson* – CEO of an operational airline with no experience (that I know of) in development

If you have a lot of clever engineers who want to add features, *put them to work improving the productivity of the testing process* instead. And if your managers or controlling partners have not done this before, prepare for the worst. Optimism, inexperience and ego precede the first crash.

The effect of *reducing rather than increasing the value* of a large, complex, risky effort is much less well known than the effect of testing. Everyone understands testing improves reliability. They only differ over how much is enough, and whether some inexpensive analysis methods can be substituted. But the ticket price reduction effect is in the category of unexpected outcomes.

We intuitively assume that if something is very valuable, we will do it very carefully. That is a misapplication of crash rate theory. The truth is if the consequences of failure are very expensive, we will do it very carefully. If the profit from succeeding is very great, we will do it very greedily, which is not cautiously, or at least we will do it under pressure to "get it done."

We also assume that if we take more time to do something, we will do it more reliably. It's true that if you rush, failure will be quick and catastrophic. But once operating at less than the rush rate, any further slowdown violates the principle that we get better and more proficient at things through repetition. Think of it like the Montana and Autobahn speed effects. Faster meant less fatalities, unexpectedly.

It is instructive to view these recommendations in terms of the two Shuttle accidents. In the first one especially, it was pressure to keep up the flight rate that was credited with causing the accident.

But wait. The original flight rate had been something like one a week, as many as our hypothetical tourism program. The flight rate had been cut like the hedge fund wanted to do, and the cost (and thus value) of each flight had escalated greatly (also like the hedge fund wanted to do). I will argue that it was the high value of each flight, aggravated by how few flights were scheduled, that made delay unacceptable. If the program had been flying every week, no one would have criticized it for delaying.

We also have to ask if due diligence was done before the flight, and we already know that the investigation found it was not. The engineers were saying not just that it might break, but that to launch in that temperature was definitely out of spec. All our conclusions about the effects of costs and flight rates should be secondary after the hard technical facts are addressed and known problems corrected.

Use operational flights as extended test flights

If your system does not "fly," then of course substitute whatever it does, possibly just operate for a day or a month like a nuclear power plant. If your process is risky, and if your R_o is not as low as you want:

- Leave the test instrumentation in place.
- Leave the engineering evaluation team in place.
- Continue to correct problems as if you were in development.
- Establish a significant budget for continuous improvement.

Leaving instrumentation in place can be combined with continuous recording via satellite of data critical to accident investigations, like flight location and black box data. Inmarsat has offered the service. All it takes is a $20 cable, but so far only Qatar Airways has signed up.[67]

Before Columbia was destroyed, foam strikes were only noticed incidentally. Afterward, additional cameras were installed for monitoring. Cameras were developed which could be attached to the manipulator arm and used to inspect for damage underneath the Shuttle. A repair kit was developed and placed on board.

There were engineers monitoring all subsystems before the accident, but they were generally powerless. After the accident there was a system engineering team organized with power to take action. Before Columbia, waivers were routinely granted. Waivers ideally should never be granted, unless there is a military emergency or natural disaster, and only for the duration of the emergency.

The corrective actions of the system engineering team post-Columbia were not, however, adequate. It could not:

- Stop operations until the foam problem was completely fixed.
- Let studies and contracts to actually fix the foam problem.
- Revert to the old foam or old tank and suffer the performance loss (even though ISS was complete and the performance was no longer needed).

As a result of that team's inability to stop operations and fix the problem, someone else stopped Shuttle operations. President Bush canceled the entire program – a much greater impact than stopping or delaying a few flights. One of the principle reasons was that it was felt that the vehicle could not be made safe because of the foam problem. Yet no engineering team was ever actually tasked with making it so, without constraining them to work within existing methods, contracts and performance specs making the task impossible. Probably just reverting to the old tank and foam would have been a significant improvement in reliability.

Mostly, adjust how you react to data. Do you react only to actual fatalities, or do you react to near misses as if they were fatalities? This is where the connection with Bayesian probability comes in. If you correctly classify the defects which didn't [yet] cause fatalities, then Bayesian updating has the same effect as the defect ratio in the crash rate equation – it implies the probability of a crash is getting higher.

But if you classify the defects as benign, which is what you formally do when you implement a waiver, then your Bayesian updating is fooled by this camouflaged data point, and updates in the wrong direction. If you don't classify these near misses as defects, they won't correctly update the crash rate formula either.

The effect of really high V_F

For the Columbia accident, foam falling off the vehicle was a known problem from the beginning, which got much worse when the tank was upgraded so that greater payloads could reach the orbit required for the ISS, an orbit necessitated by the participation of the Russians. The value of that upgrade was great. Without it, the foreign policy of the Clinton administration to maintain employment in the rocket and nuclear sector in Russia, and encourage them to evolve into a collaborative power, and not to leak nuclear material on the black market, was bankrupt. That made continued use of the new tank, which was lighter and placed greater thermal and structural loads on the foam, imperative at any cost. It was a quasi-military value. A national priority. Something you couldn't say "no" to.

Additionally, the new foam was more environmentally sound. NASA really does not have much impact on the environment, but has been crippled for decades by political pressure to avoid necessary space technology like nuclear reactors. The environmental impact of the foam was a symbol, because NASA is a symbol, and it was charged with leading the nation to a more environmentally friendly future.

Let me tell you something about manned spaceflight. It is the most difficult thing you can imagine. It is easily more difficult than curing cancer or finding immortality or solving the problems of poverty and war. If you attempt to do manned spaceflight and place any goal ahead of it, the crash rate will be unacceptable, if you even get off the ground at all.

Long-term effects

Earlier it was mentioned that over 100 or 1000 years, travel and communications became much more reliable. Now we can understand this as a combination of three effects:

- Very low V_f as costs came down orders of magnitude

- Higher C_c as lifespan lengthened and standard of living rose
- Over long periods the operating set point changed

The finely tuned reliability of air travel and communications remain, however, vulnerable to disruption by terrorists and hackers. Screening and scanning to prevent disruption strive to reduce D.

Life cycle recommendations

If we diagram our discussion in the manner of a project life cycle, we might get something like the figure below:

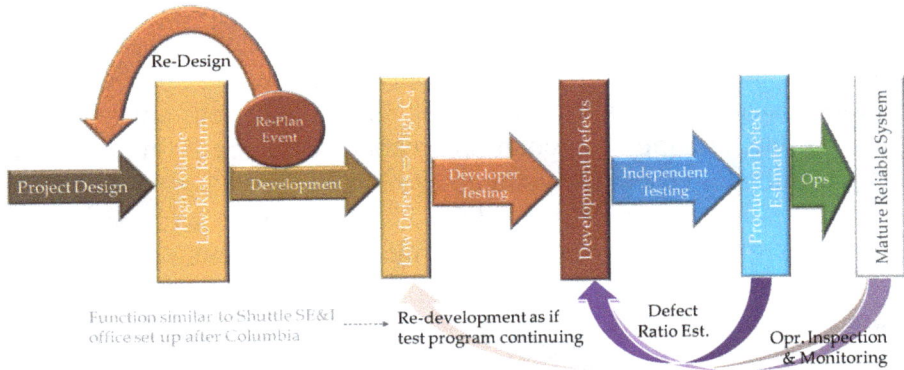

Life cycle with feedback iterations for crash rate improvement

Independent testing is used to estimate defect ratio. An independent contractor, not answerable to the project's prime contractor, performs independent tests verifying all requirements. In the process of crash rate optimization, this can help to determine an unbiased estimate of the defect ratio, or defect leakage.

In a mass-manufacturing application, the defect ratio is determined by inspecting a statistical sample. If only two vehicles are being manufactured as in our example, this concept just doesn't apply. Instead, all of the subsystems of the vehicle can be tested or inspected at intervals, and the number of defects found compared to the number of defects found during vehicle testing by the developer. The independent testing step can provide such a ratio before actual operations begin.

One objective of the testing/inspection process is to catch and correct defects at some time other than during an operational mission. This effectively reduces D, but only if the defects are meaningfully corrected. Perhaps the most important aspect of the suggested life cycle is the connection back to re-development. This implies there must be a budget for re-development, or continuous improvement. The use of new materials and techniques, or the replacement of an unreliable design with a more robust one, or one having greater performance margins, decreases the number of defects that show up during operations. In our example, we compared this new lower operational defect rate (which was the result of analysis in our pre-operational situation) to the original development-process defect rate, to show an improved defect ratio. I believe this is a valid approach based on the assumptions and derivation of the crash rate model.

To take such an approach, we assume basic commonality of the re-developed vehicle with the original one. If it is an "all new" vehicle, as is too common in the space community, then we must start over. By contrast, each new Boeing aircraft is basically a refinement of the previous one, and they can count on years of experience, improvement and denominator costs to lower their crash rate.

This is another reason I do not like going back to the capsule design. It gives up too much in evolutionary advantage. Based on public information, the lifting body or Dream Chaser type design should have been kept in the mix, and I hope Sierra Nevada will continue to pursue this. The Virgin Galactic design is not evolvable to spaceflight because the systems will have to be too different, but Dream Chaser is a spaceflight vehicle and is evolvable to a larger tourist-class vehicle.

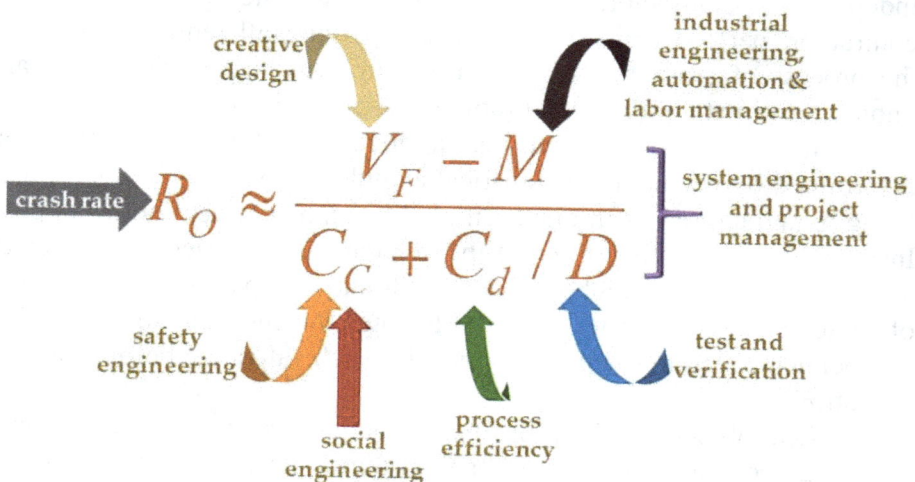

creative design

industrial engineering, automation & labor management

$$\text{crash rate} \quad R_O \approx \frac{V_F - M}{C_C + C_d / D}$$

system engineering and project management

safety engineering

test and verification

social engineering

process efficiency

Summary of Engineering Effects

Extensions

The material in this chapter was not covered in the 2013 presentation. It concerns extensions that range from simple, practical ways to estimate needed parameters, such as the defect ratio, to a well grounded method for using the theory with inhomogeneous groups rather than single entities, to a conceptual discussion of what more ambitious extensions might look like.

Estimating & using the defect ratio

We now turn to a different and simpler example, on which actual data is available, to discuss the defect ratio. The reader may have noticed that estimating the defect ratio in the previous example was a traditional risk analysis proposition, and since a deployed system was not available in large numbers for measurement, it might have involved a lot of guessing. Just what is the behavior of a real defect ratio? Most likely, it is erratic, depending on the level of scrutiny employed to find defects. Even unrelated developments may prompt consumers to use a device differently, suddenly exposing defects where there seemed to be none. In the beginning, when the project is not yet designed, it could be considered to be 100% defects (unless a functioning predecessor is available). Very little effort, in other words very low incremental (or stepwise) C_d, brings the number of defects down very rapidly. In the end, it is difficult, time consuming and expensive to find remaining defects, and every time one thinks the job is done and looks again, there are more.

The "project" we will consider now is a 300 page book. While writing the present book, I was completing another and doing final preparations for publication. There were 7 human and 2 automated editing passes for editing and proofing, and here are the data:

	cost	defects
write	$6,000	
auth proof1	$300	80
proofer	$300	60
auth proof 2	$300	50
editor	$400	30
auth proof 3	$300	40
auth proof 4	$200	20

upload	$20	3
conversion	$100	3
upload	$20	2

The "write" activity may be compared with development, and the proof activities with testing. The upload activity included some automatic checking. The conversion was manual but only looked at end notes.

What is a crash in this case? There are four candidates I can think of:

- The author is sued or threatened by an extremist group.
- The reader experiences disaster while following advice in book.
- Reader dislikes the book enough to post a bad review.
- Reader fails to recommend the book to a friend.

All of these are legitimate. Some authors are even killed over their books. Based on subject of the book, I elected not to focus on the first two. In the latter two cases, there is no direct cost impact from the "crash," so $C_c = 0$. At worst, a $100 proof copy order will be spoiled. This means all the "deterrent" in the denominator comes from C_d/D. The price was set using a tool provided by Kindle Publishing that showed maximum revenue points for this type of book, and the result was $5.49, or approximately $V_f - M = \$4$ on a per reader basis.

The question we want to ask is, can we use any of this data to estimate the defect ratio? We want to evaluate quality of the product before release, so we don't have actual data. The process of "risk analysis" does not easily apply to the book, which does not have wear and strengths of materials issues like a mechanical device, nor is it a combination of subsystems of varying function and criticality. Presumably an experienced author or editor will have a feel for how much editing is enough, but let's assume we have no clue (close to the truth). That corresponds to an innovative project designing something new for your company.

Using a spreadsheet, cumulative development costs were computed for each step, and divided by the defect ratio to get C_d. Two versions of this were examined:

- Use of D calculated as a ratio of defects at step N+1 / step N
- Use of D calculated as step N+1 / all previous defects

The first method produces a wildly varying defect ratio, as sometimes the number of defects found may go up instead of down. That may mean the testing at that step was more thorough, or it may also mean that when glaring defects were resolved, it became easier to find other defects.

The second method produces a nicely declining D as we would expect. What we tabulate below is 1/R, the reciprocal of the predicted crash rate, based on each estimate of D, for each process step, in both stepwise and cumulative versions. It reflects that one in every some number of readers will experience a "crash." Since we do not have a cost per crash for this example, the model does not distinguish between readers who simply don't recommend the book, and readers who post a bad review. As in many cases, the model gives relative information, and some external source is needed to calibrate the linear approximations, which were only intended to indicate *changes*, not absolute values.

	1/R step	1/R cum
proofer	16	16
auth proof 2	11	25
editor	14	53
auth proof 3	5	40
auth proof 4	14	91
upload	46	645
conversion	7	653
upload	10	986

We can see that the automated steps, and the steps which were not comprehensive, are not comparable to the other steps. This is a very important point. When using a technique like this to estimate, one should compare apples to apples, such as two similarly thorough passes at verification.

In the next data set we will discard the less thorough steps. In addition, two more variations are added. While these aren't "theoretically" justified, remember that this estimate has to be calibrated anyway. It is an estimate. We do not have an actual defect ratio, and cannot obtain an absolute number without calibration even if we did have, until we get actual crash data. In an established industry this is not hard to come by, but for innovative projects it is unobtainable.

The new data uses an incremental or "step" cost per defect based on just the costs for that step. This cost increases much faster than the cumulative costs, since it gets harder and harder to find defects. The chart below compares both.

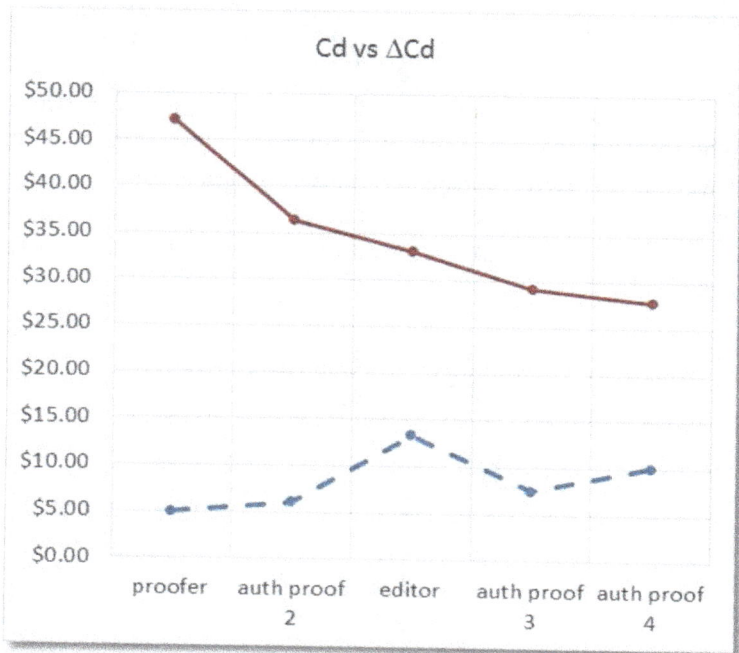

Cumulative vs. stepwise (dashed) cost per defect

The cumulative C_d generally decreases because development costs dominate, and get apportioned over more defects. But stepwise, they get harder to find and the cost each goes up.

In the following data we have either stepwise or cumulative C_d, and either stepwise or cumulative D.

	1/R step-step	1/R step-cum	1/R cum-step	1/R cum-cum	normalized
write					
auth proof 1					
proofer	2	2	16	16	1
auth proof 2	2	4	11	25	2
editor	6	21	14	53	3
auth proof 3	1	10	5	40	3
auth proof 4	5	33	14	91	6

All combinations of step & cumulative C_d/D

The stepwise versions of D, first and third data columns, appear heuristically useless to me. Either of the cumulative D columns appears to show a reasonable trend. The stepwise C_d shows a 15 to 1 improvement from beginning to end, and the cumulative Cd shows less, a 6 to 1 improvement. These are just over a factor of two different, and that is about as close as we are going to get. The more testing one does, the more the stepwise and cumulative data will converge, as testing costs become a larger percentage of total costs. More testing is good. So if you are fretting over which to use, because there is a big difference, probably you haven't done enough testing.

The final column is just the cum-cum column normalized to 1 for the first step for which there is data. Now comes the guesswork. Someone with experience will have to decide what kind of real crash rate that corresponds to. Suppose one decides it corresponds to one crash every 2 readers. Then multiply the last column by 2, and those are your estimated crash rates. The final estimate is one in 12 readers.

The knowledge as to whether one is estimating bored readers, or readers annoyed enough to post a bad review, is built into this educated guess. If the estimate was for readers simply not liking the book, a final rate of 1/12 is probably not bad. If the estimate was for readers who will post a bad review, 1/12 is probably terrible. If you plan to sell 12,000 books, that will be 1000 bad reviews. Probably no more than one or two bad reviews are acceptable.

A more valuable crash rate estimate for this project would be to estimate the ratio of 5 star, 4 star, etc. reviews. An even more valuable estimate would be to gauge the reciprocal of sales instead of crash rate, but that is not covered by our theoretical development so far.

The book was released and just a little data is available. So far there are three 5 star reviews from six readers. Better than expected. One friend mentioned liking the book but also mentioned he was keeping track of mistakes for me. That is not great, but he likes the book well enough to expend that extra effort. A crash rate of reader dislike of better than 1/6 is confirmed, and it probably is better than 1/12. It appears perhaps one or two more proofs would be helpful.

At this point it is probably not useful for the author to do another proof. Why do you think I say that? Because the author cannot see mistakes already overlooked 4 times? That might be true, but it is not the reason.

Having the author read the book again is like having the developer in charge of final testing. Anyone who has ever done this knows that what happens is, the developer sees things he or she wants to "improve." The author, speaking from personal experience, at some point introduces more errors than he removes, and that may already have begun in the fourth iteration. Interviews with successful authors sometimes reveal that they do not go back and read their old books, because it is only a temptation to revise. Revising and revising makes a good author. It makes a terrible proofreader or tester. The same is true for engineering or financial projects, and produces truisms like "*better is the enemy of good.*"

A book may seem like too-simple an example, but it contains many lessons applicable to complex projects. A book is pretty complex, after all. The reverse is true as well, i.e. applying crash rate theory to this simple project greatly improved it by justifying additional verification effort.

Lost sales and the linearity of Vf-M

Lost sales are one of the largest crash costs. Just ask any airline which has gone out of business or been restructured because of multiple crashes. We did not have an existing sales level in either the Space Tourist Transport or the book example, nor did we have a model for predicting the impact of crashes on sales, so they weren't included in C_c. But the book business provides a good example of how including them upholds the approximate linearity of crash rate with respect to the numerator, V_f-M.

Below about $3.99, readers are very forgiving of the quality of an eBook, and will accept nearly anything at 99 cents. But at the high end of that range and above, they expect an author with a national reputation who writes like it, and is backed by a publisher with a staff of editors. I make this claim from my own data, having read and reviewed around 500 books. I always read what the other reviewers say before I write my review. Reviewers keep each other honest. A one star review on a $7.99 book will often find others agreeing. A one star review on a $3.99 book will be attacked for being too harsh. On a $1.99 book, a few poor reviews will be ignored by buyers.

A $5.49 price point is the sweet spot for a typical economics or personal finance book which has around 300 pages. Above that reviewers are not especially more critical, but author revenue drops off. Since per-book revenue is increasing above that, sales are dropping fast. Therefore we conclude that to actually see the approximately linear relationship between the numerator V_f-M and crash rate, one would need to somehow quantify lost sales as a crash.

This is different than figuring the sales lost because of a crash. It is a recognition that revenue lost because of poor quality is the same as crash costs, even if there is no identifiable crash. How would we count these nebulous crashes?

We can do it with economics. If crash costs are $100k, and there is a sales baseline of $500k per year for a product, and suddenly sales drop to $300k per year without any crashes, we have an implied or virtual lost sales crash rate of:

$$(500 - 300) / 100 = 2 \text{ crashes per year}$$

If we count this decline as crashes, we will see validity of the crash rate equation over an extended range with respect to V_f and pricing variations.

The risk environment

Nothing in the theory says what the values of the terms are. Crash rate theory only allows you to predict, roughly or incrementally, your crash rate once you have determined the terms. Some of the terms can be strongly affected by other things going on in your industry or country.

There may be key features, price points or other factors which if not met, will cause most or all sales to be lost and the business to be terminated. In a business environment, this is simply competition. The competitor may well be operating at a poor choice of crash rate. But it could take time for the market to figure this out. If the competition's lack of quality is evident, it will only take a few months. In the case of automobiles, it took about two decades for brand loyalties to change.

If your company is not aware of crash rate calculations, they will be tempted to meet the competition's quality point in the short term, and perish with them in the long term. If your company is aware of crash rate, they will need a strategy to last the appropriate amount of time.

Another environmental factor, for both companies and individuals, is homogeneity, or the lack of it. Consider a mix of small and large limited liability companies. The small companies will have smaller crash costs because they can only be sued for their assets, which are small. Their investors can just open up another company. The big company has more to lose in crash costs.

I'll end where we started, with automobile fatality data, because it is interesting, and illustrates this point. Countries have fatality rates which generally are lower if they are wealthy. You can see from the figure below, the U.S. seems to be an exception.

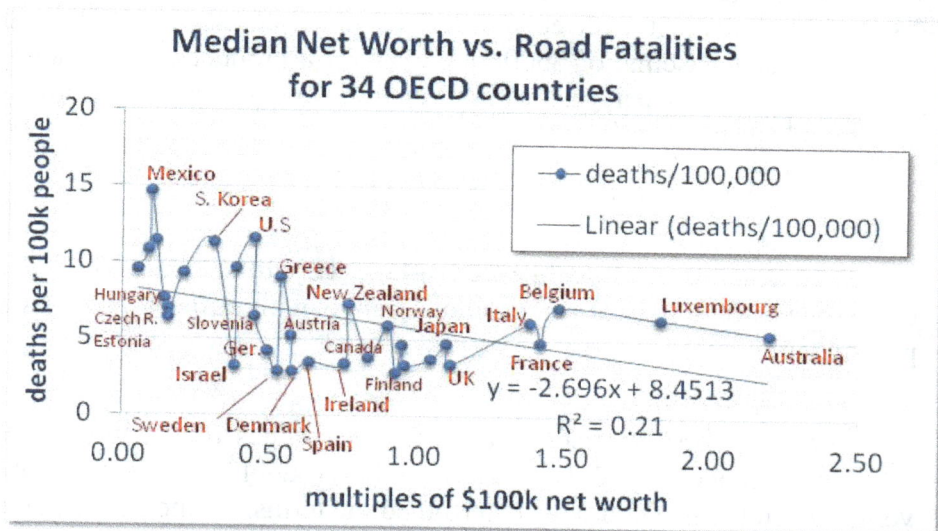

Of the 34 countries in this chart, members of the Organization for Economic Cooperation and Development (OECD), the U.S. has the 5th largest average individual net worth of $301,000, behind Switzerland, Australia, Norway and Luxembourg. Yet it has the second worst road fatality rates of the group, behind only Mexico.

However, when I was asked about this my reaction was that the "average" net worth is misleading. The U.S. has a larger proportion of poor people than most of these countries. It simply has some very rich people who skew the average. If one uses the *median* wealth, the U.S. comes in at 20th place out of 34, with half the population worth less than $44,900. The chart above plots median wealth vs. fatalities, and while there is still some dispersion, you can see that the general trend is lower fatalities with wealth, as the safety theorists expect.

This is an example of in-homogeneity. The average risk is not the same as the risk of the average. Mixing people (or companies) with different risk factors produces an overall higher risk (typically). This does not explain all mysteries, but does shed light on the road fatalities. I computed a "homogeneity" factor by dividing the median net worth by the average net worth. If they were equal it'd be 100%, very homogeneous. The U.S. factor is 15%, quite inhomogeneous. Here is the chart again with a few of the homogeneity factors noted,

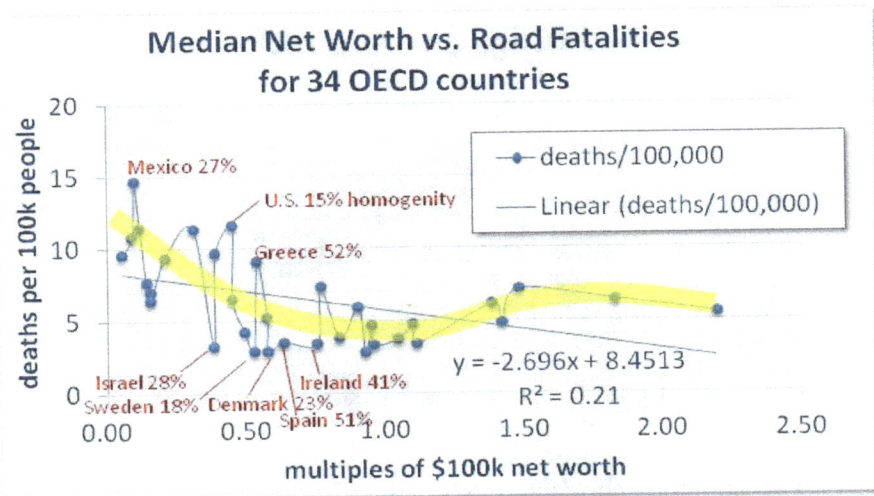

Median Net Worth vs. Road Fatalities for 34 OECD countries

y-axis: deaths per 100k people

Legend:
- deaths/100,000
- Linear (deaths/100,000)

Labels:
- Mexico 27%
- U.S. 15% homogenity
- Greece 52%
- Israel 28%
- Sweden 18%
- Denmark 23%
- Ireland 41%
- Spain 51%

$$y = -2.696x + 8.4513$$
$$R^2 = 0.21$$

x-axis: multiples of $100k net worth (0.00, 0.50, 1.00, 1.50, 2.00, 2.50)

Greece is surprisingly homogeneous for a high death rate country, but then the Greek economy is world famous, and not for stability and well-being. Generally the low death rate countries are more homogeneous than the U.S. Sweden is only barely so, but it has a huge government commitment to road safety at almost any cost, and the cultural factor of neighbors that have low death rates. In the previous version of the chart, you can see countries are often grouped by geographic region and culture.

You can work up similar data for the particular industry you engage in, to determine if the companies or customers are homogeneous, or not. You might not be in the road safety business, but you can follow this approach to refine understanding of your crash rates. In the insurance business, for example, profits are determined by a successful mix of these risks. People who have little risk, or little to lose, will not buy insurance unless forced by the government, which is true for auto and now health insurance. The people who buy it voluntarily are likely to be the ones exposed to losses. If you are in the oil business and everyone else pumps oil at an insane rate, as we discovered recently, you will too unless you are willing to give up your market share and retire. A great deal of behavior, unfortunately, is simply dictated by the behavior of others. You may be a perfectly safe driver, but if the other drivers are not safe, you will have accidents. In business, you must have confidence in your risk predictions to wait it out while a reckless competitor destroys itself, and possibly harms the entire industry.

net worth age 34-36 (2014)

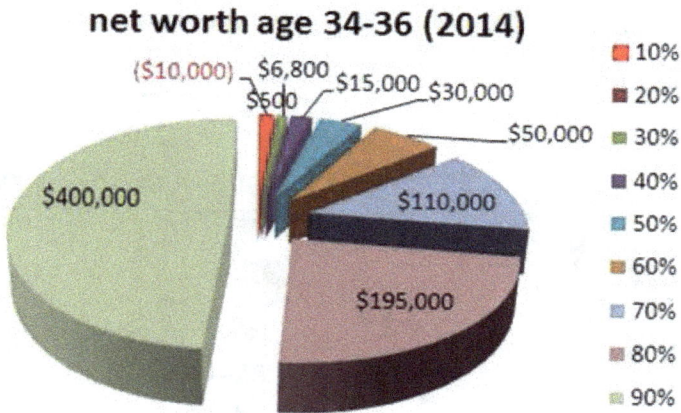

Legend: 10%, 20%, 30%, 40%, 50%, 60%, 70%, 80%, 90%

Values shown: ($10,000), $6,800, $500, $15,000, $30,000, $50,000, $110,000, $195,000, $400,000

The pie chart above shows graphically the lack of homogeneity in the U.S. population at mid career. We could look at income, or net worth, they are similar. I happen to already have this data. Below are the data in table form, showing the difference between average and median, normalized to 1 at the 50th percentile.

percentile	net worth	relative wealth
10	-10000	-0.33
20	600	0.02
30	6800	0.23
40	15000	0.50
50	30000	1.00
60	50000	1.67
70	110000	3.67
80	195000	6.50
90	400000	13.33
avg wealth 3x > median:		3.36

Now suppose there is an individual dependency of accident rate on net worth. The OECD data suggest that if net worth increases from $50k to $200k, death rate decreases from about 7 to about 3, or about ½ as fast as is net worth is changing. Approximately that means a person with half the net worth might have 1.5 times as many accidents. The results of computing the individual accident rates at each percentile point, normalized to the 50th percentile, and averaging them, are given below.

percentile	net worth	estimated relative fatalities per capita based on 1/2 wealth weighting
10	-10000	cannot process negative da
20	600	25.50
30	6800	2.71
40	15000	1.50
50	30000	1.00
60	50000	0.80
70	110000	0.64
80	195000	0.58
90	400000	0.54
average:		4.16

Using ½ weight on net worth, accident rate is 4x higher than median, equivalent to net worth of $4200

A homogeneous version of the U.S. could have a quarter of its present road fatality rate, or 11.6 / 4.16 = 2.8 fatalities per year per 100,000 people, 30% better than Germany's 4, and 6.6% better than Finland's 3. The risk environment analysis enables an economic explanation to gain considerable traction.

Approximating C_c for inhomogeneous groups

One might or might not have the sort of detailed data that is available for motorway deaths, a well-studied issue, or national wealth demographics, also highly studied. But one might have, or be able to guess, averages and medians. Let us see if we can find an approximation that does a reasonable job for the motorway deaths.

Using the U.S. wealth inhomogeneous factor (ratio of median to average) of 15%, and a wealth weighting on risk of 50% (1/2), we should come up with about a factor of 4 change in crash rate. If one's life is lost, one loses one's wealth, so we'll put wealth in the position of C_c, and we'll ignore C_d/D for now.

Let I=.15 represent the inhomogeneous factor, then the effect on crash rate is 1/I=6.7. We must reduce this using the wealth weighting, for which we'll use the symbol W. A 100% weighting would correspond to C_c x I, or the full 6.7 times the crash rate due to wealth being represented by the median rather than average value. With $C_c(I/W)$ we get 3.3 times the crash rate. The real distribution gave 4 rather than the lesser value given by using just the median. So for the U.S. traffic data, there is a fudge factor of 4 / 3.3 = 1.21 on the weighting. Let's use F=1.21, and then we have 1/(I/WF) = 4.

If we used multiple percentile points, rather than just the median, we would use the real wealth weighting factor evident from the data of W=50%. But since we want to approximate this distribution, taking into account that the really non-wealthy percentiles contribute a lot of accidents, we fudge the wealth factor a little higher at WF=60%.

There is no way to know the limits of applicability of this factor without doing many studies. In the meantime, we know that at least for the U.S., for fatality data, F=1.2 simulates the data well using just the inhomogeneous factor I, based on the median, and the wealth sensitivity factor W=50% based on the OECD data. With this we have:

$$R_o \approx \frac{V_f - M}{C_{c\_avg} I / FW + C_d / D} \tag{8}$$

Crime & war – economic vs. psychological paradigms

If we are able to apply crash rate theory to corporate competition and risk taking in automobiles, could we possibly apply it to sociological issues generally, such as war and crime?

Before even thinking about what the terms in the formula would mean for such things, or adapting them, the first thing to do in extending to a new area is to look for data on whether the economic assumption holds. And if not, what does? A good example is crime, on which there is a lot of data. But crime is in many cases the acts of individuals or small groups, and may or may not constitute a large enough percentage of the economic activity of those individuals to be subject to pressure of a specifically economic kind.

A paper in 2000 in *The Review of Economics and Statistics* looks at income inequality and crime using urban county data.[68] We might suspect a term like "income inequality" to indicate the sort of non-homogeneous wealth factors we have been using to explain motorway fatality data, and we might guess that this would increase "crashes" of some sort, and that crime incidents might be treated as crashes. We find two different cases. For property crime, inequality had no effect, but poverty did. (So did police activity.) For violent crime, poverty had little effect, but inequality did!

A similar study appeared in 1997, comparing Chicago neighborhoods.[69] An earlier study from the 1980s found lower correlation overall, but still a difference in the same direction, i.e. poverty without high inequality corresponded to a slightly lower homicide rate.[70] The conclusions of the study authors vary, but we can distill a couple of things. Violent crime doesn't pay, i.e. isn't simply due to poverty, but property crime may have an economic motive. Changes to V_f, up or down, as a motive for violent crime, were not considered by the study authors, who variously blamed it on culture or stress. These would fall into the domain of psychological factors.

In studying inhomogeneity we considered its effect on the cost of crashes. But in the case of crime, someone else's wealth, or similar attribute like prestige or power, might be the targeted payoff for the consumer (i.e. perpetrator) of the criminal act – either to seize it or destroy it. In this light, increased inequality in close proximity, i.e. inhomogeneity, makes nearby high V_f targets more accessible, increasing V_f generally, increasing the crash or crime rate.

Then why does it become violent? Consider the other half of this transaction, the target. With a lot to lose, the target may ordinarily be risk averse. But once the loss has occurred, or is in process and looks inevitable, how do the target's parameters change?

Immediately the target has less to lose, lower crash costs, because the loss has occurred or is inevitable. The target immediately accepts risky actions with higher crash rates, as a purely economic consequence. Let us assume that the target understands his or her own crash rate parameters instinctively, that is the conditions under which the "risk thermostat" may be rapidly recalibrated. Further, let us assume that the target assumes that a similar kind of risk equation governs the perpetrator. Moreover, the target expects some future "rate" of such incidents, and wishes to influence that, by providing some kind of deterrence. This might be anything from locks and fences to a bodyguard, but in many cases the target may seek to influence the future calculus of this and other perpetrators by inflicting a large "cost of crash" on the perpetrator. An ordinarily risk averse citizen may then pull out a gun and shoot someone on his or her property, even when it does not entirely seem justified, all because of this calculus. In Texas where I live it happens all the time, and apparently in Florida too. Certainly we know from what humans say and do, and our own introspection, that they behave like this in small and large ways.

So, crash rate theory *could* potentially explain relations between crime and economics that heretofore have required cultural or psychological explanations. There is a mechanism for it in the theory, as long as we do not take the environment as being fixed and immutable as the early safety engineers did before seatbelt and airbag data were available. Not only does the risk environment change, but humans make calculations based on their own risk calculations, which change the moment they lose one of their denominator terms in a crash, which are *designed* to affect the environment either directly or through the risk calculations of others. It is at least worth studying in the author's view.

What about war? Is it in fact a crime of nations? They surely use terminology like "deterrence." I have made an extensive treatment of this topic in *Money, Wealth & War*, though not addressing the crash rate equation directly. The Russian revolution began when people were losing what they had quickly, which is like the sudden loss from a crime. The Arab Spring began when a Tunisian cart vendor's cart was multiply confiscated by the police, and he set himself on fire shouting, "I just want to work!" High unemployment in Syria, sometimes reported as high as 90%, gave rise to a long revolution there and eventually the rise of ISIS. Rapid loss of power and wealth among the Iraqi Sunnis also provided fertile ground for ISIS and other insurgents. Without the pressure of the Syrian security services, the "testers" looking for "defects," i.e. rebels and opposition figures, then ISIS grew rapidly. And what was the risk calculus of those who'd join an organization that could potentially order them to their deaths? ISIS' stated goal was to form a "caliphate" with all the spoils of wealth that would ensue, including slaves and harems not even permitted in "civilized" nations.

And what of the Russian "volunteers" fighting in Ukraine? Reports are now emerging that there were two types. *Newsweek* reports that volunteers were offered from $1000 to $4000 a month,[71] which looks pretty good to young male with a child whose mechanic career dried up when laid off workers could no longer afford their cars, and whose wife makes $250 a month (he was killed after 10 weeks).[72]

Other possible extensions

This last section discusses ideas for future research, not finished methodology, or even formally derived methodology. We might use equation (8) for an inhomogeneous group of corporations having similar pricing, development and testing methods, but differing in their size and thus liability for crashes. If the other terms are also inhomogeneous, and do not have similar characteristics, we can model the other terms also, either with a detailed distribution, or with approximations as above. The C_d/D ratio might become a sum of ratios, or an approximated composite ratio equivalent to the sum.

The implication of using the crash rate equation to predict motorway fatalities takes us into a fundamentally new area, groups of individuals and the accounting of psychology in some of our parameters. The W factor which reduces the weighting of crash costs is a psychological factor. It would not be sustainable for corporations. For individuals it is sustainable because driving costs are not a self-contained and self-justified economic unit of their activity. Individuals may support uneconomic activities with income from their other activities. They may simply be taking into account vague factors, like the survival value of their freedom to move about at will, which in theory might be equivalent to some economic value, but we do not have a reasonable way at present to go about making such a judgment with numbers. Humans do it with their feelings, instincts, cultural behavior, and their "risk thermostats."

The meaning of C_d/D may be different for individuals. Is it some process they impose on themselves? Or is it imposed from without like driver training and exams, breathalyzer tests, and traffic monitoring and fines? And what are the numerator terms? Perhaps there is only V_f, which is the total cost consumers pay for transportation?

What does crash rate theory suggest about the practice of rotating managers? A literature search reveals little negative commentary, but an informal query will reveal that employees think rotating managers make short term decisions that undermine the future of organizations. They escape the crash costs C_c of such decisions.

And what of extending the idea to countries? To crashes such as war or the environment? There are several issues: Future modeling with simple equations was tried in the 1970s and at first considered very promising. I took a university class in the subject when I was a graduate student at Rice. In general the gloomy predictions didn't come to pass, and the field lost credibility. This is a cautionary tale, something we wish to avoid. An interesting hypothesis is that the predictions didn't come about *because* the predictions became widely known.

The environment is certainly an issue of knowledge. There is dispute even among those who "believe in" global warming as to the causes and appropriate quantitative models. We've seen from the war on cancer that the crash rate model is not too helpful with knowledge issues. A quantitative model of war is not established either, though the author obviously has some thoughts on the subject, having written a book on it.

We have developed the idea of a connection between the standard of living, or personal wealth, and one model parameter, the cost of crashes. While the current model is static, i.e. not evolving in time, we could create a dynamic model that might be valid as long as economic and productivity growth continue, by feeding back improvements in lifespan and standard of living into C_c, pressuring some technology into reducing D, not just reducing C_d. Incorporating specific technology changes is guesswork. In the mid-20th century, many people extrapolated spaceflight and nuclear technology. Only a few saw the turn inward toward the micro-world. Artificial intelligence technology is being extrapolated, but past extrapolations were not met.

What of short term extensions? Could we learn to predict wars three to five years in advance? The current round of wars in the Middle East (ISIS), Ukraine, etc. could be a predictable result of the 2008-9 financial crisis. World growth, even U.S. growth, has not returned to normal levels. High local unemployment preceded all those conflicts, especially the Arab Spring, for which the incitement was a cry from a Tunisian man, "*I just want to work.*" It is not current conditions but rapid change that provokes crises. C_c, in effect, is the value envisioned in three to five years.

Finally, consider the meaning of the personal C_c in small vs. large societies, and collective vs. capitalist societies. If we have a small communal society, members may see potential communal crashes as personal. In a large society we have learned that inhomogeneity in C_c may lead to a high crash rate, and suspect this may make crashes of the society itself more likely. If we try to fix this with commune-like redistribution of wealth, we may be simply reducing C_c to near zero for the individual, making crash of the society highly likely.

Indeed, the trend toward capitalism and individual responsibility, even in "communist" countries, may be an artifact of the collective calibration of our risk thermostats. But if too much inhomogeneity results, we still have a high crash rate just from size of the society and disparity between individual and societal crash costs. In the U.S. we saw the effect of a large percentage of the population at less than average wealth increased motorway death rates by 4 times, from a projected 3 per 100,000 to an actual 12 per 100,000. This disparity is driven by ownership of most production by the top 1%.

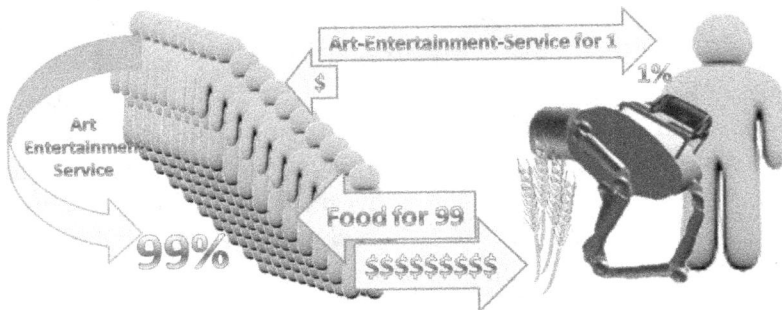

In the figure above (from *The Equity Premium Puzzle*) you can see that it is hard for the other 99% to find things of value to trade with the 1% in exchange for the production, much of which is automated. Job skills are not a reliable trade commodity in an automated economy.

With welfare, what incentive do recipients have to avoid crashes? Welfare can give them a modest standard of living, but individually they do not *lose* it when they do something foolish. So once dependent on redistributed wealth, which may be called health insurance or education grants, avoiding the stigmatized term "welfare," they have no C_c at all. It becomes zero. And if the society has no safety net and some citizens fall into poverty, we still have C_c of zero!

As a society, the actions of the crash-prone affect more than themselves. They may (will) vote for politicians who are like themselves and make foolish, impulsive choices based on the same instincts that resulted in poverty among the voters. They are likely to vote for trade, monetary and social policies that are short sighted if they have little actually to lose. But collectively the whole country may lose. That connection is harder to make in a large society.

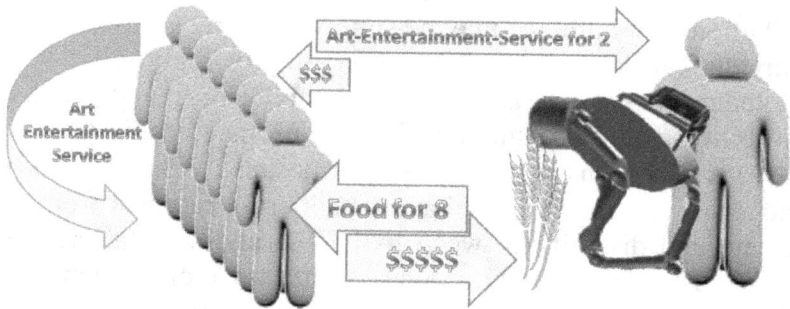

A possible way out of the trap of low C_c is to change public policy and social goals so as to make it attractive for more people, perhaps 20%, to become "owners" rather than workers. I have detailed how this might be done in *Money, Wealth & War*. The pressure toward inequality would immediately be 20 times less. As each 20% bracket yields its jobs to the 20% below, the lower 20% can be lifted into relative affluence and much higher C_c. A new bottom class may still immigrate, revealing that it is in our interests for all nations to prosper. Otherwise the planet itself suffers a low median C_c and high crash rate, exacerbated by advances in technology, productivity and interconnectedness which amount to high V_f.

Conclusion

A model has been developed that relates development process parameters to operationally deployed crash rate. This model operates at a corporate level, and provides insight into what kinds of changes might affect product reliability and safety. There is a rough analogy to risk compensation, though in detail the models address different regimes, one risk and the other economics. Process improvement and feature development are often culprits in unexpected outcomes. Testing adds to costs, but has a predictable positive outcome. Excessive value pressure is often overlooked as a culprit for complex systems, and lowering the value of function is equally often overlooked as an additional tool to reduce crash rates. The model gives good agreement with observed safety differences between auto and air travel.

Crash rate theory is not just about manned spaceflight, which serves as an example of something which is really difficult, and requires a lot of energy, control and near-perfection. We must not only "be careful," but have a quantified theory and methodology of being careful, even in the presence of unexpected effects and human behavior. If we manipulate genes, is that not an undertaking with much greater consequences for the planet than the crash of a vehicle? If we take advantage of nuclear energy and spread it across the planet, is that not also incredibly difficult to keep safe and out of the hands of terrorists? And what about dealing with the nuclear weapons that many nations possess in large numbers? Climate change? Outbreaks of new diseases? Even something as mundane as cascading financial disaster?

We discussed, in connection with the project example, the case where a company might walk away from a project if the safety requirements could not be met at a given feasible revenue level. We could have looked at reducing the passengers from ten to nine or eight, which would do much the same thing as reducing the ticket price. It might lighten the vehicle load, but would lessen the liability by reducing the number of victims, and reducing C_c increases crash rate. Walking away is always an option and we do not need a theory of crash rate for that. It goes without saying that someone must make a decision whether the objectives are worth meeting at the current time and technology level. And that leads to a final caution.

To work, crash rate theory must be followed by the true decision makers, owners, investors, or chief executives. Anyone else can and will be removed, or given arbitrary objectives, or their budget managed by line item. For example, we saw NASA budget line items zeroed by Congress.

In 1986 Richard Truly was appointed NASA's Associate Administrator for Space Flight by Ronald Reagan, to lead the recovery from the Challenger crash. Most sources agree that Truly was successful in that endeavor. In 1989 he was appointed administrator by his friend president George H. W. Bush. However, in 1992 Truly was removed. The New York Times reports: *"The White House opposed building more shuttles, wanted to entertain radical ideas for space exploration ... These and other initiatives were ... resisted by Mr. Truly and the space agency."*[73] NASA's history site says of his successor Dan Goldin: *"Despite lower budgets, his "faster, better, cheaper" approach has enabled the Agency to deliver programs of high value to the American public without sacrificing safety."*[74] However, Goldin endorsed reorganization of the space station to include the Russians, resulting in the external tank changes which contributed to the Columbia failure two years after he left, and the end of the Shuttle era.

When I related this viewpoint to an experienced project engineer involved with the Shuttle system engineering team, he remarked he had not heard or thought of it before. When I related it to one of the most experienced and effective testers in the business he asked bluntly, *"Did it work? Did the joint program with Russia prevent leaking rocket scientists and nuclear technology to other countries?"* Actually, it seems to have. The incorrect decision may not have been to upgrade the tank, but to have canceled the program. One of the key factors in Russia's current belligerent behavior in Ukraine and the Baltics is that it is not yet dependent on foreign trade, except for sales of gas which are unaffected by "sanctions." If Russia's tech sector were dependent on contracts from an international exploration program, they'd not want to be left out or left behind, and might have continued moving toward the west as they had been doing.

Take the lessons learned from attempting to make high energy vehicles fly through a vacuum with incredible accuracy, dodging cosmic rays, tolerating human error, and performing safely, and use them. Now that you have read and presumably understood this book, you have a powerful tool for analyzing and changing the rate at which huge and complicated things fail. Take it and make it better and let me know how you are using it.

End Notes

---

[1] Shuler, Robert, "Optimizing Innovation and Calamity," IJEIR, 4, 1 p50-66 (2015)
http://ijeir.org/index.php/issue?view=publication&task=show&id=410

[2] http://en.wikipedia.org/wiki/Asipu

[3] http://www.ryerson.ca/~tsly/825_history.htm

[4] "Medium sized rock cut cistern at pavuralla konda at bheemunipatnam" by Srichakra Pranav - via Wikimedia Commons

[5] Markowitz, H. (1952) Portfolio Selection, *The Journal of Finance*, **7**, 1, pp. 77-91.

[6] Peltzman, S. (1975). The Effects of Automobile Safety Regulation, *Journal of Political Economy*, 83, 4, pp. 677-725.

[7] Wilde, J. S. (1982). The Theory of Risk Homeostasis: Implications for Safety and Health, *Risk Analysis*, **2**, 4, pp. 209-225.

[8] Hedlund, J. (2000). Risky business: safety regulations, risk compensation, and individual behavior, *Injury Prevention*, **6**, 2, pp. 82-90.

[9] Stetzer, A. and Hofmann, D. A. (1996). Risk Compensation: Implications for Safety Interventions, *Organizational Behavior and Human Decision Processes*, **66**, 1, pp. 73-88.

[10] Viscusi, W. K. (2000). Corporate Risk Analysis: A Reckless Act? *Stanford Law Review*, **52**, 3, pp. 547-597.

[11] Schindler, J. D. (2003). Evaluating the Potential for Risk Compensation with the New Generation Fire Shelter, *USDA Forest Service, Technology and Development Program*.

[12] Katarelos, E. D. (2008). The Safety Cost and Its Fair Distribution to Air Transport Industry's Stakeholders, *Air Transport Research Society 12th World Conference*, Athens.

[13] Parry, I. W. H. (2004). Comparing Alternative Policies to Reduce Traffic Accidents, *Journal of Urban Economics*, **56**, 2, pp 356-368.

[14] Cummins, J. D., Weiss, M. A. and Phillips, R. D. (1999) The Incentive Effects of No Fault Automobile Insurance, *Road Safety, New Drivers, Risk, Insurance Fraud and Regulation* (G. Dionne and C. Laberge-Nadeau, eds.), Boston: Kluwer Academic Publishers.

[15] Cohen, A. and Einav, L. (2003). The Effects of Mandatory Seat Belt Laws on Driving Behavior and Traffic Fatalities, *The Review of Economics and Statistics*, **85**, 4, pp. 828-843.

[16] Carlsson, F., Johansson-Stenman, O. and Martinsson, P. (2004). Is Transport Safety More Valuable in the Air? *Journal of Risk and Uncertainty*, **28**, 2, pp. 147-163.

[17] Levitt, S. D. and Porter, J. (2001). Sample Selection in the Estimation of Air Bag and Seat Belt Effectiveness, *The Review of Economics and Statistics*, **83**, 4, pp. 603-615.

[18] http://en.wikipedia.org/wiki/German_autobahns

[19] http://en.wikipedia.org/wiki/Value_of_life

[20] http://www.slower-speeds.org.uk/files/speedkills.pdf

[21] http://en.wikipedia.org/wiki/List_of_motor_vehicle_deaths_in_U.S._by_year

[22] Derived from http://news.distractify.com/jake-heppner/astounding-facts-about-how-we-actually-spend-our-time/

[23] http://en.wikipedia.org/wiki/List_of_motor_vehicle_deaths_in_U.S._by_year

[24] http://www.complex.com/sports/2013/02/germanys-fatal-accident-rate-is-less-than-half-of-ours-despite-driving-at-155-mph

[25] http://www.biomedcentral.com/1471-2458/13/376

[26] http://www.funtrivia.com/askft/Question116086.html

[27] "Why Sweden has so few road deaths," unattributed, Feb 26th 2014, http://www.economist.com/blogs/economist-explains/2014/02/economist-explains-16

[28] "Average wage in Dominican Republic 'not even enough to eat'", unattributed, July 2014, Dominican Today, http://www.dominicantoday.com/dr/poverty/2014/7/1/51975/Average-wage-in-Dominican-Republic-not-even-enough-to-eat

[29] http://en.wikipedia.org/wiki/List_of_countries_by_traffic-related_death_rate

[30] Rudin-Brown, C. and Jamson, S., ed. (2013), *Behavioural Adaptation and Road Safety: Theory, Evidence and Action*, CRC Press, Boca Raton, FL, pp. 61-86.

[31] http://www.fiberpipe.net/~tiktin/Documents/abpr6.htm

[32] http://www.cnbc.com/id/102386596

[33] http://www.carsdirect.com/car-safety/airbag-repair-costs-prices-for-common-problems

[34] http://www.rand.org/pubs/research_briefs/RB9505/index1.html

[35] http://chriskresser.com/the-diet-heart-myth-statins-dont-save-lives-in-people-without-heart-disease

[36] http://www.lung.org/stop-smoking/tobacco-control-advocacy/states-communities/tobacco-tax.html
http://www.tobaccofreekids.org/research/factsheets/pdf/0146.pdf

[37] http://news.uic.edu/graphic-warning-labels-on-cigarette-packages-reduce-smoking-rates

[38] Spence, A. M. (1975). Monopoly, Quality and Regulation, *Bell Journal of Economics,* **6**, 2, pp. 417-429.

[39] Savage, I. (1999). Safety Regulation, Global Truck and Commercial Vehicle Technology, London: World Markets Research Centre's Business Briefing Series, pp. 86-90.

[40] Lo, Andrew W., The Adaptive Markets Hypothesis: Market Efficiency from an Evolutionary Perspective, *Journal of Portfolio Management*, Forthcoming. Available at SSRN: http://ssrn.com/abstract=602222

[41] http://traveltips.usatoday.com/air-travel-safer-car-travel-1581.html

[42] U.S. Department of Transportation (2008). 2001 National Household Travel Survey, https://ntl.custhelp.com/app/answers/detail/a_id/252/~/percentage-of-air-travel-for-business-vs-other-purposes

[43] Jet Airliner crash Data Evaluation Centre, *JADEC SAFETY RANKING 2012*, Hamburg, Germany, available at http://www.jacdec.de/jacdec_safety_ranking_2012.htm

[44] Six Sigma Online (undated). The History and Development of Six Sigma, Aveta Business Institute, http://www.sixsigmaonline.org/six-sigma-training-certification-information/articles/the-history-and-development-of-six-sigma.html

[45] iSixSigma (undated). The History of Six Sigma, http://www.isixsigma.com/new-to-six-sigma/history/history-six-sigma/

[46] Process Quality Associates (2006). The Evolution of Six Sigma, http://www.pqa.net/ProdServices/sixsigma/W06002009.html

[47] 1979 Honda Civic: http://www.uniquecarsandparts.com.au/car_spotters_guide_japan_1979.htm
1979 Chevy Nova http://www.detailshop.com/rides.php

[48] Akpose, W. (2010). A History of Six Sigma, Today's Engineer, IEEE, Dec. http://www.todaysengineer.org/2010/dec/six-sigma.asp

[49] http://blogs.scientificamerican.com/the-curious-wavefunction/2013/04/02/nuclear-power-may-have-saved-1-8-million-lives-otherwise-lost-to-fossil-fuels-may-save-up-to-7-million-more/

[50] http://www.giss.nasa.gov/research/briefs/kharecha_02/

[51] http://www.newscientist.com/article/mg20928053.600-fossil-fuels-are-far-deadlier-than-nuclear-power.html#.VN6rqCx0bEI

[52] http://blogs.ei.columbia.edu/2013/04/15/fossil-fuels-do-far-more-harm-than-nuclear-power/

[53] http://www.theguardian.com/world/2014/oct/28/japan-nuclear-power-reactors-satsumasendai-fukushima

54 http://www.uspreventiveservicestaskforce.org/Page/Document/
RecommendationStatementFinal/breast-cancer-screening
55 http://bigelowaerospace.com/
56 http://spaceflight.nasa.gov/history/station/transhab/
57 "Transhab-cutaway" by NASA -
http://web.archive.org/web/20011207025921/spaceflight.nasa.gov/gallery/images/station/transh
ab/html/s99_05363.html . Licensed under Public Domain via Wikimedia Commons -
http://commons.wikimedia.org/wiki/File:Transhab-cutaway.jpg#mediaviewer/File:Transhab-
cutaway.jpg
58 http://www.space.com/19290-private-inflatable-space-station-bigelow.html
59 http://en.wikipedia.org/wiki/SpaceShipTwo
60 "SpaceShipTwo technical diagram" Licensed under Fair use via Wikipedia -
http://en.wikipedia.org/wiki/File:SpaceShipTwo_technical_diagram.jpg#mediaviewer/File:Spa
ceShipTwo_technical_diagram.jpg
61 http://www.space.com/28203-dream-chaser-space-plane-propulsion-milestone.html
62 http://www.sncspace.com/mediakit/index.php?category=FAQ
63 http://en.wikipedia.org/wiki/NASA_X-38
64 "Dream Chaser pre-drop tests.6" by Ken Ulbrich -
http://mediaarchive.ksc.nasa.gov/detail.cfm?mediaid=66310 . Wikimedia Commons
65 http://www.uh.edu/sicsa/library/media/publications/AIAA_2013
66 http://www.reactionengines.co.uk/space_skylon.html
67 http://www.thedailybeast.com/feature2/2015/flight-370-did-not-disappear.html
68 Kelly, M., "Inequality and Crime," *The Review of Economics and Statistics*, **82**, 4
(2000). http://www.mitpressjournals.org/doi/abs/10.1162/003465300559028#.VRMQSvnF_f
0
69 Wilson, W., Daly, M., "Life expectancy, economic inequality, homicide, and
reproductive timing in Chicago neighbourhoods," *BMJ*, 1997;314:1271
70 Messner, S., "Poverty, Inequality, and the Urban Homocide Rate: Some
Unexpected Findings," *Criminology*, **20**, 1, 103-114 (1982).
http://onlinelibrary.wiley.com/doi/10.1111/j.1745-9125.1982.tb00450.x/abstract
71 http://www.newsweek.com/how-russians-are-sent-fight-ukraine-296937
72 *The Guardian*, 03/24/2015. http://www.latimes.com/world/la-fg-c1-russia-ukraine-
burying-bodies-20150324-story.html#page=1
73 http://www.nytimes.com/1992/02/13/us/nasa-chief-quits-in-policy-conflict.html
74 http://history.nasa.gov/dan_goldin.html